W9-BMX-873

THE HUNDRED GREATEST STARS

James B. Kaler

COPERNICUS BOOKS

An Imprint of Springer-Verlag

Published in the United States by Copernicus Books,
an imprint of Springer-Verlag New York, Inc.
A member of BertelsmannSpringer Science+Business Media GmbH

Copernicus Books
37 East 7th Street
New York, NY 10003
www.copernicusbooks.com

Book design by Jordan Rosenblum

Library of Congress Cataloging-in-Publication Data
Kaler, James B.
The hundred greatest stars / James B. Kaler.
p. cm.
Includes bibliographical references and index.
ISBN 0-387-95436-8 (acid-free paper)
1. Stars. I. Title.
QB801.K244 2002
523.8—dc21 2002019774

Printed in Hong Kong.
Printed on acid-free paper.
9 8 7 6 5 4 3 2 1

ISBN 0-387-95436-8 SPIN 10864993

To the community of amateur astronomers

Contents

VII Preface

IX Acknowledgments

X Introduction and Allegro

2 0 The Sun

PAGE	STAR	
4	1	Acrux
6	2	Adhara
8	3	AG Draconis
10	4	Albireo
12	5	Algol
14	6	Alpha Centauri
16	7	Alphard
18	8	Antares
20	9	Arcturus
22	10	Barnard's Star
24	11	Beta Canis Majoris
26	12	Beta Cassiopeiae
28	13	Beta Lyrae
30	14	Beta Pictoris
32	15	Betelgeuse
34	16	The Black Widow
36	17	Canopus
38	18	Capella
40	19	CD–38°245
42	20	CH Cygni
44	21	Chi Cygni
46	22	Chi Lupi
48	23	Cor Caroli
50	24	Crab Nebula
52	25	Cygnus X-1
54	26	Delta Cephei
56	27	Delta Orionis
58	28	Deneb
60	29	Deneb Kaitos
62	30	DQ Herculis
64	31	EG 129
66	32	Egg Nebula
68	33	Epsilon Aurigae
70	34	Epsilon Eridani
72	35	Epsilon Lyrae
74	36	ESO 439–26
76	37	Eta Carinae
78	38	FG Sagittae
80	39	51 Pegasi
82	40	FK Comae Berenices
84	41	40 Eridani
86	42	FU Orionis
88	43	Gamma Cassiopeiae
90	44	Gamma Draconis
92	45	Gamma-2 Velorum
94	46	GD 165B
96	47	Geminga
98	48	Gliese 229B
100	49	GRS 1915+105
102	50	HD 93129A

104 51 HZ 21
106 52 IRC +10 216
108 53 Kapteyn's Star
110 54 Kepler's Star
112 55 L 1551 IRS 5
114 56 Lambda Boötis
116 57 LB 11146
118 58 Merope
120 59 Mira
122 60 Mizar (and Alcor)
124 61 Mu Cephei
126 62 Mu Columbae
128 63 MXB 1730–335
130 64 NGC 6543
132 65 NGC 7027
134 66 Nova Cygni 1975
136 67 OH 231.8 +4.2
138 68 P Cygni
140 69 PG 1159–035
142 70 Polaris
144 71 PSR B1257+12
146 72 PSR B1913+16
148 73 R Aquarii
150 74 R Coronae Borealis
152 75 R Leporis
154 76 R Monocerotis
156 77 Rasalgethi
158 78 Rho Cassiopeiae
160 79 Rho Ophiuchi
162 80 RR Lyrae
164 81 RS Canum Venaticorum
166 82 RS Ophiuchi
168 83 SGR 1900+14
170 84 Sigma Octantis
172 85 Sirius
174 86 16 Cygni
176 87 61 Cygni
178 88 Spica
180 89 SS Cygni
182 90 SS 433
184 91 Supernova 1987A
186 92 T Tauri
188 93 Theta-1 Orionis
190 94 Thuban
192 95 Tycho's Star
194 96 Vega
196 97 VV Cephei
198 98 W Ursae Majoris
200 99 Zubenelgenubi
202 100 ZZ Ceti
204 Appendix A
206 Appendix B
208 Appendix C
210 Glossary

Preface

I have always loved the stars. I watch them, photograph them, research them, write about them. Their wonder is that they are there not simply for scientists, but for all of us, filling the night sky with their sparkling beauty. There are as many different kinds as there are stars themselves, each an individual. The heavens give us bright ones, dim ones, near ones, far ones, the aged, the young, those that help tell our ancient stories, and those nearly invisible even with the greatest of our technologies. Taken together, they relate the tale of our existence, of the birth, life, and death of the Sun on which we depend. Yet for all their scientific and cultural significance, their best role is perhaps in the peace they bring to a quiet night and to the heart.

And so one summer afternoon in 1998, with millennium fever at high pitch, with expert critics extolling their 100 favorite movies, songs, radish recipes, I thought why not *stars*? Are they not as important? Within the hour, I had about 50 that for a variety of reasons I liked best. Then the work started in earnest. I researched names within obscure categories and argued with myself about the virtues of Z Andromedae vs. AG Draconis. I added some, subtracted others, and finally—except for some minor finagling near the end—arrived at the goal: 100 stars that I really liked. They are my favorites, or at least a good selection of my favorites, and to me, *The Hundred Greatest Stars* for reasons that are both personal and scientific.

They are quite a mix. Some are included because of their positions in the sky. Others were added for their normality, several for their strangeness, still others for their history, charm, or beauty. I must admit that I cheated a bit. Famous double stars are taken together, such as Sirius B with Sirius A, giving two for the price of one. And you can hardly talk about Zubenelgenubi without bringing in Zubeneschamali, so they too are treated within one story. The Sun is not included in the 100 list, but instead leads the pack as "Star Zero."

Before describing the glories of the 100 stars, an introduction briefs the beginning stargazer on basic stellar properties and explains the astronomical terminology, without which we would be continuously tongue-tied. A separate glossary provides a quick reminder. Then we move on to the stars themselves. Each of my favorite stars is introduced by a short summary that gives the star's names, resident constellation, apparent brightness as viewed from Earth, distance, visual luminosity, and, most importantly, its significance. This quick view is enlarged by a more detailed story, which is accompanied by an image or illustration that highlights the star's character. The stars are arranged alphabetically by the name I know them by best. (Since stars commonly have two or even more names, other arrangements are possible, and these are given in Appendix A at book's end.) Each star is also cross-referenced with the others, allowing the reader to peruse the book in a non-serial fashion and reminding us that each gem is part of a larger picture. Yet each essay is self-contained, the description of one star not depending on another, allowing you to enjoy a page at random.

As the work developed, the stars began to tell a grander saga, one of stellar birth, ageing, death, and of the overall significance of the starry sky. Appendix B arranges them according to evolutionary state, beginning with their births, continuing with the lives of solar-type stars, and ending with the spectacular deaths of the most massive. Appendix C lists the stars from west to east across the sky. A couple dozen "outtakes," the stars

I listed at some point but finally removed, can be found at the publisher's website, www.copernicusbooks.com.

Are your personal favorites here? Part of the fun is to match yours with mine, to wonder why some of the outtakes did not make my main list, and to second-guess the main list: just where are Altair, Zaurak, Omega Herculis anyway? And if you have no favorites, no candidates for greatness, perhaps mine will become yours. In any case, please enjoy them and appreciate the immense range of stellar characters that Nature has to offer.

Jim Kaler

Acknowledgments

This book has been a joy to write. Perusing through the final list, I find stellar friends of long ago as well as newly-made acquaintances. *The Hundred Greatest Stars* also led me to discover both old and new stars within the publishing and scientific worlds. First thanks go to Jonathan Cobb, who instantly believed in the concept, to Jerry Lyons, who pursued it, and to Paul Farrell, who saw the book to completion. More thanks go to Anna Painter for her excellent editorial work and for assuring clarity of thought and word, to Jordan Rosenblum for his exciting design and for making sure that we had the highest quality images available, and to Mareike Paessler for her excellent proofreading and her eagle eye.

The book could not have been written without the myriad scientists who did the research, studied the stars and discovered their properties, and allowed me to report back what they have all found. Thanks to them all, and a very special thanks to the astronomers who not only allowed me to use their hard-won images, but worked to get them to me in the best form. I hope you all find that I have made good use of your knowledge and pictures. And as always, thanks, Maxine, for your support.

Introduction and Allegro

Like the night sky, this book is meant for browsing. To help guide the way, a general introduction presents basic stellar properties and processes and a primer on astronomical terminology, thus providing a solid framework for the stories that follow. Because each of the 100 stars are an intimate part of this cosmic story, they are highlighted by blue type here and throughout the text to show where these stellar wonders fit into the larger scheme.

WHAT ARE STARS?

Even though humans have been thinking about the stars for thousands of years, we still have yet to come up with a good definition of just what a star *is*. The simplest description is that stars are the self-luminous condensates of the fragmented dusty gases that fill interstellar space. With some distinctive exceptions, they can also be described as massive balls of gas that are squeezed down and heated under the force of their own gravity until the interior temperatures and densities are so high that they are stabilized and kept from contracting by some outward force. Usually, the stabilizing push comes from thermonuclear fusion, the process by which lighter atoms are joined together under conditions of great heat and pressure to form heavier atoms, releasing energy we eventually see as starlight.

Most stars fuse hydrogen (the dominant stellar building block) into helium, but the fusion of heavier atoms (helium into carbon and so on) can also do the job. Stars that are just being born, however, have not yet begun the fusion process, while "dead stars" have stopped and are either stabilized by something else or by nothing at all. So the most inclusive definition of a star is *a body that is undergoing thermonuclear fusion, or that has spent or will spend part of its lifetime in such a state.*

THE GREAT FUSION ENGINE

Like our planet, stars are constructed from all the chemical elements, from hydrogen to uranium and even beyond. Each element is made of a different kind of atom, which consists of even tinier protons, neutrons, and electrons. Much like the Earth and the Sun are held together by gravity, positive protons and negative electrons are linked together through electromagnetic force. While neutrons carry no electromagnetic force, like protons, they generate the powerfully attractive (but short-range) "strong force," which allows them to attach tightly to each other and build atomic nuclei that are surrounded by families of electrons. The kind of chemical element depends on the number of protons in the nucleus, 1 for hydrogen, 2 for helium, 26 for iron, and so on.

Nature makes about 100 different kinds of elements. The nucleus of ordinary hydrogen is one plain proton, but if the proton joins a neutron partner, the combination (or "isotope") is called hydrogen-2, or deuterium. Ordinary helium, helium-4, has two protons and two neutrons; a one-neutron isotope, helium-3, is also found in Nature. Most lighter atoms have a range of stable isotopes, but if there are too many or too few neutrons, the atom becomes unstable and falls apart, releasing radiation. While the simplest and lightest atoms, hydrogen and helium, are rare on Earth, they totally dominate stars. The Sun is 92 percent hydrogen and 8 percent helium, as are most other stars. All the other known chemical elements occupy a measly 0.15 percent. In general, the heavier the element, the less there is of it.

While the chemical elements may seem independent, they can be transmuted back and forth into one another. Heavy radioactive elements release the strong force by breaking apart, or fissioning, while lighter ones release it by fusing together. Both processes help generate iron and nickel, and this is the reason that there is so much of these two elements in the Universe. Stellar fusion can take place only where the heat and pressure are so great that protons can slam together, overcoming their natural electrical repulsion and sticking together via the strong force. The great gravitational compression found in the deep cores of stars provides the neccessary conditions. In the center of the Sun, the temperature is 16 million degrees Celsius, and the density is 150 times that of water. From the center out to about a quarter of the solar radius, protons sequentially hammer together to make helium in what is known as the "proton-proton chain." As the protons fuse, they release a small amount of the strong force, which is converted to high-energy radiation and eventually to beautiful starlight.

THE NIGHT WATCH

While the great majority of stars are quietly fusing hydrogen into helium, many others are in various states of aging, a process that is loosely (and not very accurately) called "stellar evolution." To understand it, let the Sun set, watch the sky darken, and observe.

But First We Have to Get Organized

Among the first things learned by nascent stargazers are the constellations—named patterns of stars. Every culture has invented them, and all have seen the sky differently. As far as we know, "our" constellations were originally named by the ancient Babylonians and Sumerians about 2000 BC and were later adopted by the early Greeks, handed to the Romans and to the Arabs of the Middle Ages, and finally passed to us: 48 classical figures set against the nightly sky.

Constellations, of course, are not real, but are imaginative representations of things that were important to the cultures that invented them. We see sacred symbolism in the dozen constellations of the Zodiac, through which the Sun, Moon, and planets seem to move. Elsewhere, the sky is populated by mythical heroes and demons that were (and still are) used to tell great stories, tales about Orion the Hunter or the Great and Small Bears, Ursa Major and Ursa Minor. Given that the constellations are made from more-or-less random distributions of stars at different distances, few look like what they are supposed to be: they represent, not portray.

The old constellations only occupy those parts of the northern sky that are populated by bright stars. Vast, unnamed areas of the night aky appear between these ancient constellations, of course, and to the naked eye, appear to contain only drab stars, or no stars at all. Then there are the stars of the far Southern Hemisphere, which were not accessible to the Greeks and their predecessors. The intellectual revolution of the European renaissance prodded astronomers of the day to fill the celestial sphere with fanciful figures of their own. Hence, we find "modern" constellations that represent industrial artifacts (a furnace, a microscope, a telescope, and so on) as well as additional animals. The final list includes 88 formal constellations, some big, some small. Added to these are a vast number of informal figures such as the Big Dipper, a part of Ursa Major.

Though some astronomers disdain the constellations as mere mythology, which has no place in a scientific world, they remain important in piquing initial interest and, equally, in organizing the sky, allowing us to recognize and name the stars. Hundreds of stars are identified by proper names (mostly derived from Arabic and Greek) that have something to do with the stars' appearances or placements. DENEB, at the tail of Cygnus the Swan, means just that, "tail," in Arabic.

Such names are still widely used for the brighter stars, but they are difficult to recall, and duplication can cause confusion (there are several variations on "Deneb"). A more logical system was needed. Around 1600, Johannes Bayer attached Greek letters to the brighter stars within the constellations, beginning with Alpha and, if enough stars were present, ending with Omega. Later, others applied the letters to the stars in the growing list of modern constellations. Thus, there are 84 "Alpha" stars, one for nearly each formal figure. (Four constellations do not have them.) The basic rule is to order the stars within a constellation by brightness, "Alpha" being the brightest star, and so on, but there are more exceptions to the rule than adherents. Bayer also named stars based on their position within a constellation and on other rules known only to himself. As a result of Bayer's work, for example, SIRIUS is also known as "Alpha of Canis Major." Since Latin was the educated language of the time, the constellations have Latin names. To imply possession, Latin uses a case ending: "of Canis Major" is indicated by "Canis Majoris," "of Lyra" by "Lyrae," and so on. Sirius is Alpha Canis Majoris, and Rigel is Beta Orionis. All good astronomers thus need to know, or at least recognize, the 88 constellations plus the 88 possessives. Based on the observations made by the late-seventeenth-century English astronomer John Flamsteed, stars are also numbered sequentially from west to east within a constellation. VEGA (Alpha Lyrae) is also 3 Lyrae, and 61 CYGNI is the star whose distance was first measured. When no Greek letter is available in a constellation, astronomers commonly use these Flamsteed numbers.

For faint stars, the constellations fail, so we use coordinates and catalogue numbers. Like the surface of our planet, the sky can be fixed with a coordinate system. On Earth, we measure position by latitude in degrees north or south of the equator and longitude east or west of the prime meridian. In the sky, we use the analogous "declination" and "right ascension." Declination measures stellar position north or south of a celestial equator that lies above the Earth's equator, and right ascension measures the angle between the star and a circle that runs from the sky's apparent north rotation pole through the vernal equinox in Pisces (where we find the Sun on the first day of Northern-Hemisphere spring). Every star has a pair of such coordinates and can be placed on a map. Catalogues are built upon these positions, giving us the ability to name any star we wish. But the process still begins with the friendly constellations.

Distance

Of the stellar properties needed to address the all-important evolutionary processes, none stands out more than distance. Distances allow us to find the crucial stellar luminosities and masses, which help us understand stellar lifecycles. Using the distances to stars, we can also construct a picture of our astronomical environment.

The huge distances encountered require appropriate units. The Sun, the closest of all stars, is 150 million kilometers away, a fundamental distance in astronomy called the "Astronomical Unit," or AU. Other stellar distances are so great that we use a much longer unit, the "light year," or the distance a ray of light will travel in a year at a speed of 300,000 kilometers per second. Given 31 million seconds in a year, the light year contains 63,240 AU.

The closest star—ALPHA CENTAURI—is 4.4 light years (270,000 AU) away. Most of the naked-eye stars lie within 100 light years of us, but some are up to thousands of light years distant. Using these measurements in conjunction with others (such as radio observations of interstellar gases), we find that all the stars you see at night are part of our Galaxy (its common name, hence the upper-case "G"). Shaped like a flat, thin plate with a bulge in the middle, and containing 200 billion stars, our Galaxy stretches 80,000 light years from one side to the other. Our Sun is offset by 25,000 light years from the center. The Sun lies in the disk, and the disk's stars encircle us. Their combined light makes the broad, white band we call the Milky Way. In a dark sky, the Milky Way is astonishingly bright, especially in the direction of its center in Sagittarius. Surrounding the disk is a huge spherical halo only sparsely occupied by stars.

The disk rotates. Each star has its own separate orbit, not unlike the planets of our Solar System. In the vastly bigger Galaxy, however, it takes 200 million years for the Sun to go all the way around. As a result of differing orbits, the stars slide past each other, the nearby (or faster) ones appearing to move across the sky relative to those in the more-distant background. Over time, the familiar constellations will disappear as the stars shift away from one another.

The distances of stars, based on the AU, are determined principally by parallax. To visualize this concept, move your head from side to side, and nearby objects in the room will appear to shift back and forth through an angle that depends on their distances from you. As the Earth goes around the Sun, the closer stars appear to shift in the same way. The technical "parallax" of a star is half the angle through which it appears to shift over the course of a year. It is also the angular radius of the Earth's orbit as seen from the star. The distance in "parsecs," the professional unit, is the inverse of the parallax in seconds of arc (1/3600 of a degree) and equals 206,265 AU. One light year of 63,240 AU translates into 0.307 parsecs.

The current practical limit to parallax measure is around 1000 light years, beyond which we use a variety of indirect measures that are ultimately calibrated on parallax. The distances to each of the 100 stars presented here are given in light years, sometimes very approximately—astronomy is not always the precise science it is touted to be.

Magnitude

In the second century BC, the Greek astronomer Hipparchus divided the stars into six brightness categories now called "magnitudes." The first magnitude stars were the brightest, while sixth was the faintest he could see. The system, still in use today, is now placed on an exact mathematical scale, wherein five magnitudes correspond to a factor (ratio) of 100 in actual brightness. A first magnitude star is thus 100 times brighter than a sixth magnitude star, so that each magnitude unit is 2.51. . . times brighter than the next one down. The exact scaling caused the very brightest stars to climb to magnitude zero, even "minus first." (The Sun, because it is so close to us and therefore intensely bright, is –27th!) Magnitudes are decimalized and are routinely measured to 0.01 of a unit, or to even greater precision. The "first magnitude" category includes stars between magnitude 0.51 and 1.50, "second magnitude" from 1.51 to 2.50, and so on.

With telescopes, we can go much fainter and see to higher magnitude numbers. A backyard telescope allows us to see to 10th to 15th magnitude, while the Hubble Space Telescope can approach magnitude 30—stars that are 4 billion times fainter than the human eye can see alone. Since these magnitudes are as seen from Earth, they are properly called "apparent magnitudes."

Color

There is, however, a complication: our eyes do not see all the radiation a star produces. Stellar radiation is not confined to "light." Light behaves, in part, as an electromagnetic wave, and the color you see depends on wavelength, the distance between the wave crests. Of the colors of the rainbow—the spectrum—red has the longest waves, violet the shortest. The light we see has an average wavelength of five hundred-thousandths of a centimeter. The wavelengths of the invisible infrared are longer than those of red light, while radio wavelengths are longer yet. Shorter than violet is the ultraviolet, which is followed by X-rays and by gamma rays, which have wavelengths billions of times shorter than the rainbow's colors. The shorter the wavelength, the more energy the wave carries. Radio waves are benign, while X-rays and gamma rays are deadly. Gamma rays are generated by thermonuclear fusion. Fortunately, they do not escape directly from the stellar cores, but are degraded by their passage through the stars' outer layers (in a process of continuous absorption and re-emission) into benign light.

Though most stars are gaseous throughout, they can still be said to have "surfaces" at which the gases become highly opaque (rather like a cumulus cloud). Depending on the kind of star, these surfaces have a range of temperatures. Stellar temperatures are always given on the Kelvin scale. If you remove all the heat from a body, it drops to –273 degrees Celsius. Since there is no longer any heat energy, the body can get no colder, and we have reached "absolute zero." To avoid negatives, the Kelvin scale (abbreviated to "K") is simply degrees Celsius + 273. Stellar surfaces range from around 1000 K to an extreme of 1 million K. The hotter the star, the greater its ability to emit energetic radiation. Consequently, cool stars are red and hot stars are blue, in keeping with the energies of spectral colors. To an astronomer, "color" is a handy reference to temperature. Colors actually seen by eye are subtle, and the cause of sometimes vociferous, if not futile, arguments. The coolest stars are so chilly that they radiate only at longer wavelengths, in the infrared, rendering them invisible to the eye. Hot stars, on the other hand, radiate all across the spectrum, from the infrared into the ultraviolet and X-ray. By virtue of its heat alone, a cool star cannot produce high-energy radiation, but a hot star can, and does, produce low-energy radiation in addition to the high-energy kind. A variety of processes unrelated to temperature can cause stars of all persuasions to radiate, even in the X-ray and gamma-ray spectral domains.

Different astronomical detectors (the eye, photographic films, photoelectric devices, all outfitted with different-colored filters) "see" the stars differently. Whereas the eye may see a red star brighter than a blue star, photography will record the blue star as the brighter. At the extreme, a very cool star that can be wonderfully bright to an infrared detector can be completely invisible optically. Since Hipparchus's apparent magnitudes are as seen with the eye, they are called "apparent visual magnitudes," and are the ones used in the stories that follow.

Luminosity

Just as a distant flashlight looks fainter than a nearby one, the apparent magnitude of a star depends on distance. We can compare stars only if we know their true luminosities, the amounts of energy that they radiate into space. To compensate for differing distances, astronomers use "absolute magnitudes," mathematically pushing or pulling the stars to a standard distance of 10 parsecs, or 32.6 light years. Once stars are lined up at a common distance, we can begin to appreciate the astounding range of stellar properties. Absolute magnitudes range from around –10 (for normal hydrogen-fusing stars) to fainter than +20. Our apparently brilliant Sun shrinks to an absolute magnitude of 4.83. Since each set of 5 magnitudes is a factor of 100 in brightness, 15 magnitudes represents a luminosity ratio of 1 million. That is, the most luminous stars are 1 million times brighter than the Sun, and the dimmest are 1 million times fainter. And this range does not include the exploders ("supernovae") and the "substars," which are too small and cool to fuse much of anything.

However, to understand stars, we need total power, not just that seen at visual wavelengths. Visual magnitude alone gives no idea of how much invisible infrared or ultraviolet radiation may be packed into starlight, nor does it give much information on red, blue, or violet. Fortunately, the stellar temperature, which is determinable from the star's color or from the details of its spectrum, can be used to calculate the amount of "missing" radiation and correct the visual magnitude. The adjustment is called the "bolometric correction," and the resulting corrected magnitudes are "bolometric magnitudes." Since for solar-type stars most radiation already comes out in the visual, the corrections are rather small. The absolute bolometric magnitude of the Sun is 4.75, not much different from its absolute visual magnitude. For the temperature extremes, however, they can be huge—many magnitudes.

Once bolometric magnitudes are known, they can be converted into true luminosities in solar units, that is, expressed as "so many times" the solar luminosity. To place stellar power in familiar terms, the Sun's luminosity is 3.86 million-trillion-trillion watts, of which the Earth captures a mere tenth of a billionth.

KINDS OF STARS

Bulk properties alone—luminosity and color—give little idea of what the stars are actually like. We can begin to get an idea by correlating properties such as absolute magnitude and color, the latter directly related to temperature. While these methods are helpful, there is a better way: through the detailed examination of stellar spectra. Astronomers, simple folk, use a system of but 12 letters to classify the majority of stars based on their spectra—nine to indicate temperature class (from high to low, OBAFGKMLT), and another three (R, N, and S) to indicate chemical variations. Superimposed on these are other categories that give additional compositional variations and the star's "luminosity class," which is linked to the state of the star's evolution.

Temperature Class

A star's temperature class, both historically and in modern practice, involves the observation and analysis of its spectrum. To see a spectrum, we use a prism or other device to spread starlight into a rainbow. If you stretch it far and cleanly enough, you will see myriad thin dark lines set perpendicular to the flow of color.

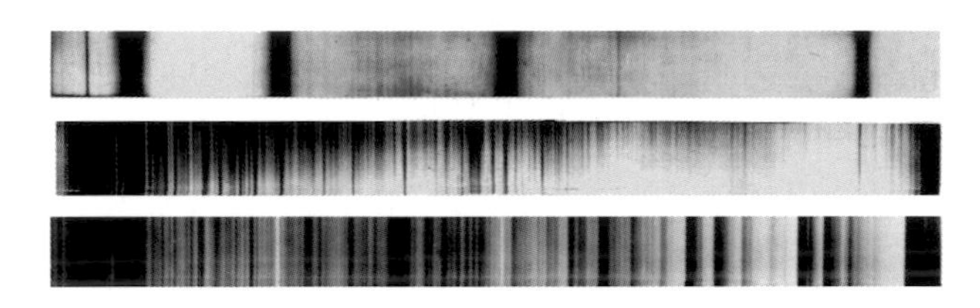

Spread out the light of the Sun or any other star, and it will break into a rainbow of colors. Spread it out with great refinement of color, and dark "lines" will appear, each one produced by absorption of starlight at a specific wavelength by a specific atom or ion. Stars are classified by the appearance of the spectrum. At the top is a class A star that displays powerful absorptions of hydrogen. In the middle is a class G star like the Sun with myriad absorption lines of ionized and neutral metals. At the bottom is an M star with its strong molecular bands, particularly those of titanium oxide. (*University of Michigan Observatory.*)

These lines are created as escaping starlight is absorbed by atoms and ions that lie in the star's outer layers.

Each kind of atom has its own set of narrow, dark absorptions, or "absorption lines," that are created by the atom's electrons. In its normal state, every atom is surrounded by a cloud of negative electrons equal in number to the number of nuclear protons, rendering the atom electrically neutral. If enough energy is supplied to an atom, an electron can change its energy and can even be torn away. The atom is then positively charged, which makes it into an ion. If two electrons are stripped off, the atom is "doubly ionized," and so on.

Stars consist of a mixture of atoms and ions. The higher the temperature, the more ions there are. Electrons absorb and release energy at specific wavelengths (that is, colors) that depend critically on how the electron cloud is structured. Change the number of electrons, and you change the absorption pattern, or the "absorption spectrum." As a result, ions have completely different absorption spectra than their parent atoms. Indeed, every ion of every atom has a different set of absorptions, allowing us to identify the chemical makeup of any star (using laboratory comparisons). Hundreds of thousands of absorption lines have been identified.

Stars have a striking array of absorption spectra, which were first explored in the mid-nineteenth century. Some stars are dominated by amazingly dark and broad hydrogen absorptions. Others have only weak hydrogen, but display prominent metal lines. Still others have no hydrogen at all, showing off almost hopelessly complex sets of molecular absorptions. To begin to understand these discoveries, astronomers first had to group the stars—by using their spectra—into similar categories.

The classification scheme used today was developed over 100 years ago at the Harvard College Observatory. Stars were initially ordered alphabetically based on the strengths of their hydrogen absorption lines. (The "strength" of an absorption line refers to the amount of energy that the line removes from the spectrum. "Strong" lines are wide and dark, while "weak" lines may be no more than slight depressions in spectral brightness.) Stars with the strongest hydrogen absorption were called class A, next down B, and so on. After much trial and error and rearrangement to achieve smooth transitions in line strengths from one class to the next, the result was the classic seven-letter "spectral sequence," OBAFGKM, which encompasses most visible stars. The spectrum of class O is characterized by ionized and neutral helium (along with hydrogen), while class B is marked by hydrogen and neutral helium, and class A by very strong hydrogen. Toward the cooler classes, we see the development of lesser metal ions and neutral species. The Sun (class G) displays strong absorptions of ionized calcium and neutral sodium. Class M is characterized by molecules, particularly titanium oxide.

The spectral sequence was quickly seen to correlate with color: O stars are blue, while M stars are red. Since blue stars are hot and red ones cool, the spectral sequence must also correlate with temperature. The sequence stretches from around 50,000 K at class O, through 5800 K at class G (our Sun), to 2000 K at class M. The spectral sequence is not a chemical sequence, as it first appeared, but a temperature sequence that involves molecular destruction and atomic ionization.

At lower surface temperatures (from roughly 4000 K down), molecules, which are rather fragile, can form and remain stable. As temperature increases, the molecules move faster and collide with other atoms and molecules, reducing them to neutral atoms. If you increase the temperature and batter the atoms more forcefully, electrons are kicked away, and ions begin to form. As the temperature goes up farther, double-ions come onto the scene, then triple ions. Each chemical element, however, proceeds at its own pace, so that at any temperature there is a great mix of neutral atoms (or even molecules) and of different kinds of ions. Not until rather high temperatures, over 10,000 K, are hydrogen ions created, and not until even higher temperatures, around 20,000 K, do helium ions appear.

Even for a given ion, the strengths of the absorptions are very temperature-dependent. At cooler temperatures, absorption of light by hydrogen is so inefficient that hydrogen lines are not seen at all. As the temperature climbs, their strengths dramatically increase to class A. At this point, hydrogen ionizes and its absorptions become weaker again. (The hydrogen ion, a bare proton, has no spectrum, because it has no electron to absorb energy.) Once these principles were understood, astronomers found that the stars of the standard sequence were all made out of the same stuff: about 90 percent hydrogen, 10 percent helium, and a salting (0.2 percent) of all the other atoms.

In 1999, two more "cool" letters (L and T, which respectively exhibit metallic hydrides and methane) were added after astronomers discovered new kinds of ultracool infrared-radiating stars. The final (as of now) descending temperature order is OBAFGKMLT. The temperature sequence of stars is continuous, with each class covering a range of temperatures. Class A, for example, extends from 7000 K to 9500 K, and scientists

CLASS	COLOR	TEMPERATURE	SPECTRAL CHARACTERISTICS
O	Blue-white	32,000–50,000	Ionized helium; neutral helium and weak hydrogen lines are also visible.
B	Blue-white	10,000–30,000	Neutral helium, hydrogen lines strengthening toward the cooler subtypes.
A	White	7200–9500	Very strong hydrogen decreasing toward cooler subtypes as ionized calcium increases.
F	Yellow-white	6000-–7000	Ionized calcium continuing to increase in strength and hydrogen weakening. Lines of neutral elements strengthen.
G	Yellow	5300–5900	Ionized calcium very strong, hydrogen weaker. Neutral metal lines, particularly of iron and calcium are prominent.
K	Orange	4000–5200	Strong neutral metallic lines; molecular bands of CH and CN (cyanogen) become prominent.
M	Orange-red	2000–3900	Strong absorption bands of titanium oxide and large numbers of metallic lines.
L	Red-infrared	1500–2000	Metallic hydrides and alkali metals prominent.
T	Infrared	1000	Methane bands prominent.

The Spectral Sequence.

found that they needed a finer scale. Today, each of the classes is decimalized, class A running from A0 at the hot end to A9 at the cool end and merging with F0, which runs to F9, and so on. On this finer scale, the Sun is a G2 star.

Luminosity Class

Temperature and its related spectral class by themselves are insufficient to describe the observed array of stars. Within any spectral class, there is a range of luminosity that increases as the temperature cools. The large majority of stars are called dwarfs, a severe misrepresentation because these are all normal stars like the Sun, and there is nothing small about them. The dwarfs form a "main sequence" of stars ("dwarf" and "main sequence" are synonymous). At the hot end, we find the O stars, which are all very luminous. At the extreme, which starts at O3, these stars can radiate nearly 1 million times the power of the Sun, their radii 20 times solar. As temperature declines along the main sequence through classes B and A, the luminosity drops dramatically. At the hot end of class A, the luminosity is just 50 times that of the Sun. M, L, and T stars are remarkably dim, with none visible to the naked eye; stars of class T radiate under one hundred-thousandth the solar power. As we will see later, the main sequence is really a hydrogen-fusing mass sequence that runs from about 100 solar masses to below one tenth solar at the bottom.

Dwarfs are called such to distinguish them from a much more luminous and larger set of stars called giants, which are found in classes O through M (classes L and T are confined exclusively to the dwarfs). The differences separating the dwarfs from the giants are by far the biggest in the cooler classes, class M and K giants singled out as red giants. "Giant" is no misnomer. The amount of energy emitted per square meter of surface area by a solid or by a hot, high-pressure gas increases quickly along with temperature. Double the temperature and the surface emits 16 times as much radiation. To be bright, cool stars must be large in radius. An extreme M giant can have a total, or bolometric, luminosity 20 magnitudes (100 million times) greater than an M dwarf. At their biggest, they can have diameters over 100 times that of the Sun, and, if placed within our Solar System, could stretch across the orbit not just of Earth, but of Mars. Toward hotter classes, the difference between giants and dwarfs becomes progressively less, and near class A the two more-or-less merge.

Dramatically larger and brighter than giants are the supergiants. Within class M, these can be hundreds of times more luminous than run-of-the-mill giants. The largest would nearly span the orbit of Saturn, and in volume could hold over 1 billion suns. While supergiants span the whole range of spectral classes, like giants they too become progressively smaller toward the hotter end of the scale. On top of the supergiants, at the very pinnacle of stellar greatness, are extremely luminous hypergiants.

Smaller stars are just as important. Between the giants and the dwarfs in both luminosity and radius are the subgiants, while just a bit less luminous than the dwarfs (for the same temperature) are the subdwarfs. Nothing seems "normal," does it?

At the bottom of the traditional size scale are stars not much bigger than Earth. The first of these found looked white and were given the name "white dwarf" to distinguish them from ordinary dwarfs like the Sun. The most famed is SIRIUS's orbiting companion. The term is a misnomer, as white dwarfs also span the temperature scale from over 100,000 K to just over 3000 K, giving us white dwarfs that are blue and those that are red. All are united by their tiny size.

Instead of all these terms, astronomers frequently use Roman numerals to infer luminosity. At the top, Roman I, are the supergiants. From there we descend through the bright giants (II), giants (III), subgiants (IV), to the dwarfs (V). Supergiants have such a wide range, they are divided into Ia for the brighter and Ib for the dimmer. These are appended after the spectral type to give a near-complete stellar description. The Sun is a G2 V star, ANTARES M2 Iab (in between Ia and Ib), ARCTURUS a K1 III giant. Long after these appellations were first applied, the rare hypergiants came on the scene, and since there is no "zero" in Roman numerals, were designated "0." Do not confuse the O ("O") of the spectral sequence with the zero ("0") of luminosity.

All these relationships are best seen in a graph of luminosity plotted against spectral class called an "HR diagram," named after Danish astronomer Ejnar Hertzsprung and the American Henry Norris Russell (see page XX). There are yet many more kinds of stars, oddballs that cannot be placed on the HR diagram.

Return now to the night sky. The daytime class G Sun seems bright, yet even at a rather short distance of 100 light years, it would not be visible without telescopic aid. Neither obviously would be the less-luminous stars, including the white dwarfs. The naked-eye sky is, therefore, naturally populated by the intrinsically brighter stars. The constellations are

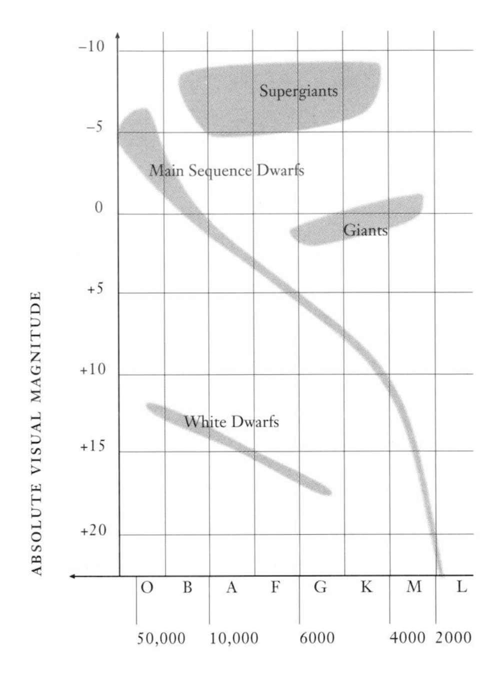

The HR Diagram. The dwarfs of the main sequence cascade downward in luminosity from class O through the whole spectral sequence (OBAFGKML, T left out) from absolute visual magnitude –7 or so to fainter than +20. When their ultraviolet light is accounted for, the O dwarfs are much more luminous than they appear here. To the right are the giants and supergiants (both of which actually merge with the dwarfs), and down at the lower left are the dim white dwarfs. Hypergiants lie at the top fringe of the supergiants. The main sequence is a zone of stable stars that fuse ordinary hydrogen (hydrogen-1) in their cores. It is also a mass sequence that runs from around 100 solar masses at the top down to the minimum of 0.08 solar (and continuing into the lower-mass brown dwarfs of class L, those that begin fusion with deuterium, hydrogen-2). Subgiants, giants, supergiants, and white dwarfs all represent various stages in the evolution of the stars as they approach death. (*Kaler, James B. 2001.* The Little Book of Stars. *New York: Copernicus Books.*)

largely composed of class A and B dwarfs and giants (and even include a few supergiants). The real count is terribly different. By far the most numerous kinds of stars are the faint M, L, and T dwarfs, which constitute at least 80 percent of the whole tally. As one climbs the main sequence, the numbers quickly drop; fewer than one star in a million is class O.

Chemistry

The chemical compositions of the stars of the main sequence are similar to the solar composition: about 92 percent hydrogen, 8 percent helium, and 0.15 percent everything else. Within the stellar "everything else," oxygen tops the list, followed by carbon, neon, and nitrogen. By contrast, the Earth, nearly devoid of hydrogen and helium, is made from this "everything else." In the Sun, however, there is but 1 atom of oxygen for every 1300 hydrogen atoms. The Earth is a third iron, nearly all of it in the central core, while the Sun has just 1 iron atom for every 32,000 of hydrogen. And iron atoms are populous compared with others, uranium having under one trillionth the abundance of hydrogen.

As a general rule, the differences along the spectral sequence among main sequence dwarfs are caused by temperature. That rule changes among the other luminosity categories—the non-dwarfs—where variations in chemical composition can affect the spectral sequence. Most giants have twice as much oxygen as carbon. However, a few reverse the ratio to become "carbon stars," deep red stars whose spectra are instantly recognizable by absorption bands of carbon compounds. (Don't let the name fool you; hydrogen still dominates.) Such giant stars were assigned the letters R and N. Class R has temperatures that parallel the G and K giants, while class N temperatures parallel those of class M. Both classes are now commonly lumped together as class C (for carbon). Between class M giants (in which oxygen exceeds carbon) and class N (in which carbon exceeds oxygen) lie the giants of class S, in which carbon about equals oxygen. S stars are characterized by absorption bands of zirconium oxide rather than titanium oxide. These odd stars, both S and N, are enriched in a variety of other chemical elements as well, including the zirconium that helps form the S stars.

Various other kinds of stars display off-beat chemical compositions but carry no formal spectral letters. Among the giants and supergiants, we find different stellar sets in which hydrogen is nearly absent, leaving them very rich in helium and in carbon, nitrogen, and other elements. (These are not to be confused with the above "classical" carbon stars—they are very rare and very different.) Subdwarfs, which for a given temperature are a bit fainter than ordinary dwarfs, can be very low in elements heavier than helium.

The various kinds of stars—displayed in the different zones of the HR diagram—are understood in terms of stellar evolution. Main sequence dwarfs fuse common hydrogen in their cores, which makes them very stable and long-lived. Stars of the middle main sequence turn into giants and then into white dwarfs, while the class O and hotter class B stars become supergiants and eventually explode.

To this picture, we add the chemical evolution of some stars and the chemical evolution of the entire Galaxy. As stars age, they create the heavy atoms from carbon to uranium and spew them back into the interstellar gases, from which new stars are born. As a result, the heavy-element content of the Galaxy increases with time. Old stars thus have a lower metallic content, and low-metal subdwarfs are all old. They are visitors from outside the Galaxy's disk (and as such are given away by high velocities), some even from the extended Galactic halo that is estimated to be 13 or so billion years old. Before further exploring these evolutionary pathways, however, we must look at the most important of all stellar properties, mass, which is determined from one of Nature's great gifts: double stars.

Mass

Stars love to double-up, triple-up, multiple-up. Doubles and multiples are charming to see in the telescope and have a strong presence in the 100 list. Most multiples are arranged in laddered hierarchies. In a triple star, a distant companion orbits a closer double. Or two close doubles will orbit each other, and then maybe a fifth star (or another double) will orbit the double-double. True multiples, in which all the stars mutually orbit each other, do exist, but are not stable. (The best-known is Orion's Trapezium, which contains THETA-1 ORIONIS.) As the numbers increase, they become star clusters. Ragged open clusters dominate the Galaxy's disk, while populous spherical globular clusters are found in the Galaxy's outer halo.

The most important feature of double stars (or "binaries") is that they allow us to determine stellar masses. Nothing is more important. All of stellar evolution revolves around mass. The components of binary stars are in mutual orbit around each other. Some members are tens of thousands of AU apart, while others effectively touch, whirling around each other in hours. The orbits of double stars are described by Isaac Newton's laws of motion and gravity. For a given separation, Newton showed that the orbital period depends on the combined stellar masses. Since orbits are always mutual (each star going about the other), the relative sizes of the orbits provide the ratio of the masses. The combination of the sum and the ratio gives the individual masses of the stars in relation to the solar mass.

Doubles come in three traditional (though highly overlapping) classes. The components of visual binaries can be observed directly through the telescope, allowing us to watch the pair swinging around a common center of mass. Spectroscopic binaries, on the other hand, are usually too close together to be seen as individuals. But being close, they have short periods and high orbital velocities. The motion of a star along the line of sight is detectable by the Doppler effect, which causes the wavelengths of light to appear shorter for an approaching body and longer for a receding one. The Doppler effect is detected in stars through slight wavelength shifts in the absorption lines. In the ideal case, a closely-spaced double star will produce two observed sets of absorptions that alternately shift back and forth as the stars orbit. The shifts yield velocities, orbital characteristics, and, again, information on stellar masses.

Unfortunately, there is no simple way of knowing how a spectroscopic binary's orbital plane may be tilted to the line of sight. As a result, we can get only lower limits to both the orbital velocities and the masses. However, if the orbital plane lies in (or near enough to) the line of sight, the stars may eclipse each other. From the duration of eclipses, these binaries give us information on stellar dimensions as well as additional data on the orbit and on stellar mass.

Binaries encompass all the different kinds of stars—the dwarfs of the main sequence, giants, supergiants, white dwarfs, and so on. Consequently, we can study the masses of the different varieties and correlate stellar mass with class and luminosity. The collection of studies shows that the main sequence is really a mass sequence. The greater the mass, the greater the internal temperatures, and the greater the stellar luminosity. Luminous O dwarfs top the list at up to 100 Suns. Masses drop below 10 solar in class B, then to a few solar in A and F. Class M stars drop below half a solar mass. From class F on down, dwarfs fuse their hydrogen to helium via the direct proton-proton chain. Hotter stars conduct fusion through the "carbon cycle." By sequentially absorbing protons, carbon turns into isotopes of nitrogen and oxygen, and then, with the ejection of helium, drops back to carbon. The carbon cycle also powers the nuclear engines of giants and supergiants and can severely alter stellar nitrogen and carbon abundances.

As we descend in luminosity through class M and into class L, the dwarfs encounter something of a magic number, 0.08 solar masses. Below this limit, the mass is too low to generate enough internal heat to run the full proton-proton chain. Thus, stars of lesser mass cannot fuse their hydrogen. Although they can still fuse their natural deuterium (hydrogen with one neutron attached) into helium, they are primarily powered by gravitational contraction. Such stars are called "brown dwarfs." These stars are mixed in with real dwarfs in class L, and are the exclusive population of class T (where they are sometimes known as "methane brown dwarfs"). The lower limit of the formation of such stars is unknown. The low-mass limit for any kind of fusion is one-eightieth the mass of the Sun, or 13 times the mass of Jupiter. At lower mass, the definition of a star becomes murky. Can

Nature make stars directly from interstellar clouds that are as small as planets (which accumulate from grains in dusty disks around new stars)? Can Nature make planets that are so massive they start fusing their hydrogen and behave like stars? Nobody knows.

The mass-luminosity relation seen among the dwarfs does not apply to the other categories of stars (though giants do tend strongly to be more massive than the Sun, and supergiants to be nearly as massive as O dwarfs). Giants and supergiants are "too bright" for their masses (as compared to dwarfs), while white dwarfs are too dim. Observations of white dwarfs in double-star systems show them to have masses comparable to that of the Sun, most falling between 0.5 and nearly 1.5 solar masses, all packed into a body typically the dimension of Earth. The result is an average density of 1 million grams per cubic centimeter—a metric ton in a sugar cube. The origin of all the divergences from the dwarf's mass-luminosity relation is found in the processes that govern stellar evolution.

STELLAR EVOLUTION

Except in rare instances, we cannot see individual stars aging. Instead, we see different stars that lie at distinct points along a variety of evolutionary pathways. As fusion changes the chemical composition of the core, the structure of the entire star must respond and change. We use the laws of physics to calculate how the internal changes affect the star's outside appearance, namely its surface temperature and luminosity. When we do—behold!—we explain all the sky's variety of stars. The results depend on mass. Here is what happens.

Low Mass

The rates of nuclear fusion are terribly sensitive to temperature. Though higher-mass stars have much more hydrogen fuel available to them, their internal temperatures and densities are higher, and they use their fuel supplies at much faster rates. As a result, they die far more quickly than low-mass stars. A 100-solar-mass star will use all of its core hydrogen in only 2 million years, whereas a feebly-radiating class M red dwarf will live for trillions. If a cluster is born with an intact main sequence (O through L), the dwarfs will die from the top down, O stars first, then B stars, and so on.

Since we can calculate how long it takes for each spectral class to die, we can date a star cluster by noting the most massive dwarfs it still contains. By finding the oldest clusters, we can date the Galaxy itself! The ancient globular clusters, which are filled with sub-dwarfs, have no main sequence dwarfs hotter than about spectral class G8 (0.8 of a solar mass), and thus come in at about 13 billion years. Stars cooler than G8 take longer than the age of the Galaxy—indeed of the Universe—to die, and therefore not one of them ever has (one of the reasons that there are so many).

Middle Mass

The middle of the dwarf sequence runs from 0.8 solar masses up to about 10 solar, covering everything from class G through nearly all of class B. The stars of this middle range change their external characteristics only slightly as they use their internal fuel. A 10-solar-mass B1 star will last for 20 million years, a "Sun" for 10 billion. In spite of these large differences, however, the internal processes are much the same.

The main sequence, or dwarf, stage is terminated by the end of core hydrogen fusion ("burning" in astrospeak). As the nuclear fire is quenched, the now-helium core shrinks under the force of gravity, and the release of held-back gravitational energy causes the core to heat. The increased temperature fires up fresh hydrogen in a shell around the inert, contracting helium core. So much extra energy is released that the star begins to brighten and to expand, the expansion causing the surface to cool. As the central fire dies out, the star becomes first a transitional subgiant and then a full-blown red giant of class M. A star like the Sun will become 1000 times brighter and will expand to a size greater than the orbit of Mercury.

At a temperature of 100 million K, the quiet helium in the core takes on new life. The conditions are so extreme that three helium atoms can combine to form carbon and gamma rays; adding another helium atom makes oxygen. The released energy stops the core's contraction and provides a measure of stability. The star settles down for about 1 billion years as a smaller, warmer (and less red) giant that is fusing helium in a core surrounded by a hydrogen-burning shell. Many such stars flock the nighttime sky; most (the class K giants) are recognizable by their rather orange colors.

When the core helium finally runs out, the carbon-oxygen core shrinks and heats. The star grows even brighter and larger than before as it "climbs the giant branch" for the second time, now with a shell of burning helium nested inside one of burning hydrogen. The two switch on and off in sequence, the fusion slowly adding to the bulk of the carbon-oxygen core. Some of these "second-ascent" giants, the more massive ones, are as big as the orbit of Mars and radiate well over 10,000 solar luminosities. As they reach their peaks, these stars become unstable and begin to pulsate, changing their radii, surface temperatures, and—most dramatically—their apparent magnitudes. The second-ascent star MIRA, for example, is for a time an important part of its constellation, and then completely disappears from view, its cycle taking about one year.

But there is far more to these stars than mere size. The Sun, like most other stars, sheds mass through a "wind." In spite of its impact on the Solar System (creating aurorae and blowing comet tails), the solar wind is so soft that, at the present rate of mass loss, it would take the Sun 10 trillion years to evaporate. Stellar winds are driven by the pressure of radiation and controlled by magnetic fields. As a star expands, its surface gravity decreases, which, with the increased luminosity, causes the wind to blow ever more fiercely. On the second ascent of the giant branch, the evaporation time would be only a few hundred thousand years, short enough for the star to effectively evaporate itself right down to, or near, the nuclear burning zone.

Huge numbers of non-energy-generating nuclear reactions also take place within the nuclear-burning regions. As the stellar ejecta depart into space, by-products that may have been brought up by convection can escape. A gift to the cosmos, new stars will be made from this enriched matter. Much of the carbon and nitrogen on Earth, and a variety of other elements (strontium, zirconium, molybdenum), were made in such stars before the Sun and Earth were born. The departing ejecta also begin to reveal the hot carbon core, which, in turn, ionizes and lights up the fleeing mass. For a short time, the core is surrounded by a lovely glowing, expanding shell misleadingly called a "planetary nebula." Instead of absorption lines superimposed on a continuum, planetary nebulae produce the opposite—emission lines—at the same wavelengths at which the absorptions occur. The thin nebular gas is heated and ionized by the hot inner star, and when the electrons get back together with atomic nuclei, the electrons lose energy rather than absorb it; other emission-producing mechanisms operate as well. Analysis of the emissions clearly shows newly-created helium, nitrogen, and carbon being washed back into space.

The nebulae last for about 100,000 years, after which time the gas is so dispersed that it can no longer be seen. What remains is the shrunken core, a new white dwarf made mostly of carbon and oxygen, whose only fate is to cool forever. The Galaxy, even our immediate surroundings, is filled with ancient white dwarfs, all too faint to be seen by the unaided eye. The Sun, reduced to around half its current mass, will be one some day.

High Mass

O stars are different. These high-mass stars, really starting at about class B1, cover the range from 10 to over 100 solar masses. As the birth masses of "middle-mass" stars increase, so do the masses of the final white dwarfs. White dwarfs have no internal nuclear burning to keep themselves supported. Their support comes instead from the outward pressure provided by their free electrons (those liberated from atoms by ionization). Electrons can behave like tiny waves, and their wave-like nature constrains the number of electrons of a given speed that can be packed into a small enough volume, one comparable to the size of the atom itself. If you can pack no more, you stabilize the star.

But a funny thing happens when the star's birthweight reaches about 10 solar masses. At this point, the white dwarf that grows within the star reaches another magic number, a mass 1.4 times that of the Sun (the "Chandrasekhar Limit," after Subrahmanyan Chandrasekhar, who won a Nobel Prize for his work). At this mass, the density and temperature are so great that the electrons lose their ability to support the star. Beyond 10 solar masses, white dwarfs are impossible, and the star's only recourse is to explode.

At first, high-mass O stars behave much like those of middle mass. As they begin to die, they develop helium cores that shrink and then fire up to burn to carbon and oxygen. But instead of becoming giants, they become supergiants; and the very highest-mass stars turn into hypergiants. Like their lower-mass cousins, high-mass stars also erode themselves through winds, though in a much grander fashion. If enough matter is lost, their surface chemical compositions can change dramatically as freshly-made carbon and nitrogen are dredged up by convection into thinning hydrogen envelopes. Unlike the stars of middle mass, the core, subject to vastly greater compression, now carries on. When helium burning finally creates the carbon-oxygen core, the core shrinks and heats. When the next critical temperature is reached, the core begins to fuse a mixture of neon, magnesium, and oxygen. When carbon-burning stops, the core shrinks and heats yet again, eventually fusing to a mixture of silicon and sulfur. The silicon-sulfur core repeats this behavior and finally fuses to iron, each successive stage going faster.

Iron cannot fuse to anything and generate energy, and as a result, it becomes unstable. The problems start when a star has grown about 1.5 solar masses worth of core iron that is the size of Earth. Teetering briefly on the edge, the core will come crashing down to the size of Manhattan well within a second of time. The iron breaks back into its constituents, into protons and neutrons, and the positive protons join with the free negative electrons to create yet more neutrons. When the ball of neutrons hits a density around 100 million metric tons per cubic centimeter, it stabilizes for the same reason the white dwarfs stabilize, except because of the wave-nature of the neutrons rather than of the electrons.

The core, having hit "bottom," violently bounces, and—aided by forces we are only beginning to understand—sends a powerful shock through the outer parts of the star. The shock blows the rest of the star apart in a grand and brilliant supernova that can reach an absolute visual magnitude of –18, well over 1000 times brighter than the most luminous stable star. In the extraordinary heat, nuclear reactions within the expanding layers run amok and create all the known chemical elements. These elements are then blasted into space and mixed with the fresh elements made by less-massive stars. A typical core-collapse supernova can, for example, generate a tenth of a solar mass of iron.

The exploded debris, expanding at tens of thousands of kilometers per second, produces a glowing cloud, a supernova remnant that eventually converts into a glowing shock wave within the surrounding interstellar gases. At its core, too dense to be destroyed, lies the neutron star. As the iron core collapses, the growing neutron star's rotation rate naturally increases, resulting in a tiny body that can spin dozens of times per second. The star's magnetic field, concentrated by the great increase in density to over 1 trillion times that of Earth, aligns itself at an angle to the rotation axis. Radiation beamed along the magnetic axis wobbles in space like a berserk garden sprinkler. If the Earth happens to lie along the surface of the cone of energy generated by the spinning neutron star, we receive a blast of energy, and our astronomers announce the discovery of another "pulsar." More than 1000 are known. When young, a pulsar radiates over the whole spectrum, from radio to gamma ray. Over time, the neutron star radiates away its spin energy and rotates more slowly. The pulsar eventually emits only low-energy radio waves and finally disappears from "sight" altogether.

Like the white dwarfs' Chandrasekhar Limit of 1.4 solar masses, neutron stars, held up by their neutrons, have a similar, though less-well-known growth limit of 2 or 3 solar masses. If the supernova produces a core in excess of the neutron-star limit, it must collapse, but this time forever. As the radius decreases, the gravity increases. Eventually the gravitational field becomes so powerful that the velocity needed for a body to escape—that is, not ever to return—hits the speed of light. Light cannot get out, and the body disappears from view, collapsing inside the surface at which it was last seen, or its "event horizon." The gravity of this black hole remains the same. We just can no longer "see" it.

The story becomes more complex when the mutual evolution of double stars is considered. If two stars are close together, they can transfer matter back and forth between them. The interactions become especially strong as the stars, in turn, expand during the courses of their evolution. The results can be spectacular. They begin with weird effects caused by matter that streams from one star to the other. The effects may continue as fresh hydrogen drawn from a companion onto a white dwarf or neutron star erupts in a nuclear-fusion runaway. At the extreme, a white dwarf can be so overloaded by a companion's gaseous gift that it exceeds its maximum mass limit, explodes, and annihilates itself in a supernova greater than one produced by core collapse.

THE CYCLE OF LIFE

Wander out on a dark clear night and look at the Milky Way, the disk of our Galaxy. It is not a smooth stream, but is filled with dark splotches, rifts, and holes. These are not absences of stars, but dark clouds of dusty gas, the birthplace of stars, seen in silhouette against the background. The space between the stars is filled with a lumpy, shredded interstellar medium made of both gas and tiny dust grains. The grains were initially condensed out of, and launched from, giant and supergiant winds (silicates from oxygen-rich giants, carbon grains from carbon-rich stars). They were then highly modified as they absorbed ambient metal-loaded gases and became coated and embedded with ices. While interstellar gas is heated by starlight, in the thicker clouds, the dust absorbs the heating light and chills the gas. Dense knots within the clouds, perhaps compressed by supernova blast waves, are forced to condense under their own gravity. As they shrink to become new stars, they rotate faster and faster and develop surrounding, rotating, dusty gaseous disks. They would spin themselves apart were it not for braking mechanisms that include the effect of the Galaxy's magnetic field (which grabs onto a spinning blob) and the formation of jets that pour out perpendicular to the disks. In many instances, the forming stars do indeed spin themselves apart to become double and multiple stars.

Unless the circumstellar disks are somehow disrupted by neighbors, the dust within them can accumulate into larger solid bodies, a process that produced the Sun's planets 4.5 billion years ago. In the inner Solar System, where it was hot with sunlight, the growing bodies could not accrete light gases and molecular ices, and they became metallic, rocky planets like Earth. In the colder outer System, however, the planets could grow fat on the light raw material to become Jupiter and Saturn. Farther out yet, where the disk began to thin, it could only create smaller bodies—Uranus, Neptune, and near the edge, little Pluto. And we are back home watching the Sun, a star of middling mass, slowly age eventually to become a giant; we look outward to the other stars and see all the processes happening, from star formation to explosions. We see the whole picture, the stars continuing to be born, to live, to die, in the process of death creating new life—the heart of the story to be given in the 100 tales that follow.

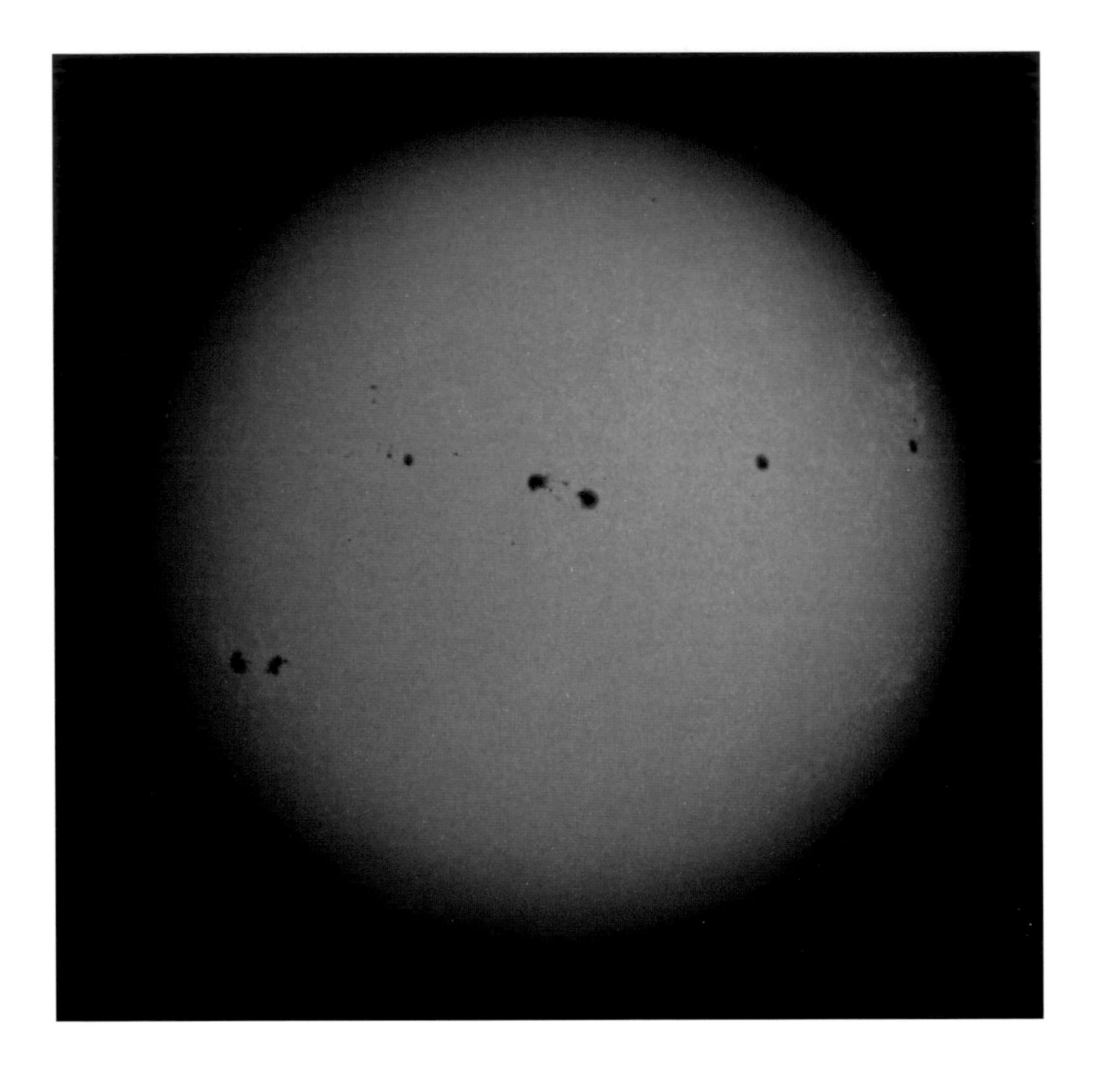

The magnetically spotted Sun, turning once a month, provides nearly all our heat and energy and is the base with which to understand all the other stars that surround us. *(NOAO/AURA/NSF.)*

Residence:	THE ZODIAC (ARIES THROUGH PISCES)
Other name:	IT NEEDS NONE
Class:	G2 DWARF
Visual magnitude:	–26.75
Distance:	1 AU = 0.000016 LIGHT YEARS = 7.9 LIGHT MINUTES
Absolute visual magnitude:	4.82
Significance:	OUR STAR.

THE SUN

Universally revered as the giver of warmth and life, we hardly think of the Sun as a star, a surmise that goes back to ancient times. It is different because it is our star, the one we can see most closely (though at no time attempt to view the Sun, as it can burn the eye), and the one we know most about.

With the Sun only 150 million kilometers (1 Astronomical Unit, or AU) away, astronomers can detect amazing detail. The next-nearest star of similar brightness, **ALPHA CENTAURI**, is 271,000 times farther away. The Sun, an astronomical yardstick, is the reference to which all other stars are compared, as their properties are commonly expressed in solar units.

The Sun is 1.4 million kilometers across, the equivalent of 109 Earths set side by side, and has a mass of 2 million-trillion-trillion kilograms, or 330,000 Earths. Most astonishing is its luminosity of 400 trillion-trillion watts. This immense energy, pouring from a body with a yellow-white "surface" (actually an opaque gas) at 5777 K, is generated by thermonuclear fusion in the Sun's deep core, which comprises about half its mass. The core's temperature reaches nearly 16 million K and has a density 14 times that of lead. No matter the density, the core is, like the rest of the Sun, entirely gaseous.

The outer layers, which are convectively unstable and continuously seethe with up- and down-welling gases, are made of 92 percent hydrogen, 8 percent helium, and 0.15 percent of everything else. Oxygen dominates the "everything else," followed by carbon, neon, and nitrogen. In the heat and pressure of the core, the hydrogen is slowly turned into helium, while a small amount of mass is lost and converted to energy in the process. After 4.5 billion years, the core of the Sun is now about half helium, and there is enough hydrogen left to last for another 5 or so billion years. When the hydrogen runs out, the Sun will become an orange giant like **ARCTURUS** and so many of the stars we can see with the naked eye.

The Sun spins slowly with a period of 25 days at its equator, and the spin and churning outer gases produce a magnetic field about five times the strength of Earth's. The rotation wraps the field into powerful ropes with strengths thousands of times greater than the Earth's, which break through the surface and, by inhibiting convection, chill local areas to make sunspots. The magnetism heats a thin outer layer, the corona, to 2 million K. The corona's thinness makes it dim and visible to us only during a total eclipse, when the bright surface is blotted out by the Moon. A thin "wind," controlled by magnetism and luminosity, flows from the surface of the Sun and blows past the Earth, making comet tails point away from the Sun. Bubbles in the solar wind produced by collapsing solar magnetic fields also make the northern and southern lights. Since many stars exhibit similar phenomena, the Sun provides a way to understand stellar behavior. The stars, in turn, help us understand our own, precious, personal star.

Crux, the famed Southern Cross, lying in the Milky Way and anchored by Acrux, leads first-magnitude Beta and ALPHA CENTAURI (the bright star at far left) across the southern sky. *(© Akira Fujii.)*

Residence:	CRUX
Other name:	ALPHA CRUCIS
Class:	B0.5 SUBGIANT + B1 DWARF
Visual magnitude:	0.77 COMBINED
Distance:	320 LIGHT YEARS
Absolute visual magnitude:	–4.19 (COMBINED)
Significance:	A BEAUTIFUL BLUE DOUBLE STAR.

ACRUX

Aside from the constellations of the Zodiac, one modern figure, Crux, the Southern Cross, is perhaps the most famed. Stolen from Centaurus and epitomizing deep southern skies, this brilliant four-star figure is one of the very few honored in music (in the lovely tango by Richard Rogers, "Beneath the Southern Cross," from the documentary *Victory at Sea*). And epitomizing Crux is the brightest of its stars, the Alpha star, which is the 13th-brightest in the sky. This most southerly of first magnitude stars is clearly visible only south of the Tropic of Cancer. From nearly all of the temperate Southern Hemisphere, the star is circumpolar, never setting. Too far south to have been named by the inhabitants of the ancient Middle East, the star is so significant that it received a modern name, Acrux, from Alpha Crucis.

Acrux is more than what meets the eye. It is not one star but two, made of near twins separated by 1.44 seconds of arc. The brighter, Alpha-1, is a first magnitude star. By itself, it would rank 20th in the sky. The other, Alpha-2, is a bright second magnitude (1.73). Both are hot blue class B (almost class O) stars, with temperatures of 28,000 and 26,000 K. The brighter is classed as an evolving subgiant. The dimmer one is a hydrogen-fusing dwarf. Their common distance of 320 light years reveals enormous respective luminosities of 25,000 and 16,000 solar.

And still surprises await. While Alpha-2 is a single star, Alpha-1 again is double, but one whose components can be detected only through their spectra. As they orbit, the stars move back and forth along the line of sight, causing their absorption lines to shift cyclically as a result of the Doppler effect. The components of Alpha-1, thought to be around 14 and 10 times the mass of the Sun, orbit each other in only 76 days at a distance of about 1 AU. On the other hand, Alpha-1 and Alpha-2 (which has a mass of around 13 solar), orbit over such a long period of time that the motion can hardly be seen. Given their minimum separation of 430 AU, the period is at least 1500 years, and perhaps much longer. A fourth star, another class B subgiant, lies 90 seconds of arc away from triple Acrux. It shares Acrux's motion through space and at first appears to be gravitationally bound to Acrux. However, if at Acrux's distance, it is under-luminous for its class, and is probably just a line-of-sight partner over twice as far away. The masses of Alpha-2 and the brighter component of Alpha-1 suggest that the stars will someday explode. However, the fainter component of Alpha-1 may escape to become a massive white dwarf.

Best of all, Acrux leads our vision to one of the most glorious regions of the Milky Way, the great band of starlight formed by the disk of our Galaxy, in which the Sun is embedded. To the east of the Cross lies the "Coalsack," a dense cloud of star-forming interstellar gas and dust that blocks the background starlight. The Coalsack is so dark and prominent that the Incas made a constellation of it, calling it "Yutu," a partridge-like southern bird. Northeast of Acrux is one of the finest star clusters in the sky, the "Jewel Box," so bright that it was named after the tenth Greek letter name, Kappa Crucis. Once seen, Acrux and its surroundings can never be forgotten.

Brilliant SIRIUS so dominates Canis Major (the Greater Dog) that bright Adhara (see also #17), at the lower right of the triangle down and to the left of Sirius, is often forgotten. *(© Akira Fujii/DMI.)*

Residence:	CANIS MAJOR
Other name:	EPSILON CANIS MAJORIS
Class:	SINGLE BLUE B2 BRIGHT GIANT
Visual magnitude:	1.51
Distance:	425 LIGHT YEARS
Absolute visual magnitude:	–4.07
Significance:	THE BRIGHTEST SECOND MAGNITUDE STAR AND THE BRIGHTEST OF ALL IN THE ULTRAVIOLET.

ADHARA

Winter in the Northern Hemisphere brings an extraordinary display of celestial jewelry, of sparkling blue-white diamonds, their turbulent twinkling intensified by the deep cold, Orion's belt contrasting smartly with reddish **BETELGEUSE**. Followed by his faithful companion, Canis Major, the Greater Dog, is jeweled as well, led by brilliant white **SIRIUS**, which rides high above a striking triangle of still other gems. Chief among these is Adhara, the brightest of the second magnitude stars, which gives the Dog two fine luminaries.

Bright as it is, Adhara was still given the modest Epsilon (fifth Greek letter) designation by the great namester Johannes Bayer, probably because of its southerly position as the western foot of the Great Dog. Having nothing to do with the mythological Greek dog, the Arabic name means "the Virgins." Adhara is part of an old Arabic constellation of the same name composed of a bright triangle and another pair of nearby stars.

Though we might ignore Adhara in favor of its flashier northern neighbor, Sirius, Adhara is by far the grander, as it is still nearly first magnitude, even though it is 50 times farther away. From its distance and apparent magnitude, the star is visually 3700 times brighter than the Sun, and when the ultraviolet radiation from its 18,000-K surface is factored in, Adhara's luminosity increases to 15,000 solar, implying a mass about a dozen times solar. The radius, calculated from temperature and luminosity, is about 13 times that of the Sun, in keeping with that found from its tiny angular diameter.

Adhara's brilliance and spectrum show that it has left the main sequence of hydrogen-fusing (dwarf) stars behind and has begun to die. The star probably has a quiet, shrinking helium core, one that will eventually fire up and fuse to carbon. Its high temperature, relative proximity, and evolutionary status combine to make Adhara the brightest ultraviolet source—as seen from Earth—in the sky. That is, if you had ultraviolet eyes, Adhara would appear to be the sky's brightest star. Things, as will so often be seen, are not always as they seem.

Adhara's significance, however, lies not so much in the star itself, but in its line of sight. The so-called space between the stars is in fact filled with a rich mixture of dust and gases which are a combination of the original matter of the Galaxy and of the leavings of stars. Much of the thin gas of interstellar space is visible as a result of the spectrum lines that it superimposes on the spectra of the stars, particularly in the ultraviolet. Here, the great luminosities and simple spectra of the blue class B stars provide a great background of radiation for their study. By examining these stars, we find that the gas of interstellar space is depleted of many heavy elements as they are condensed onto dust grains.

Adhara is a primary source of the ultraviolet radiation from hot stars that ionizes, or strips electrons from, the atoms of the local interstellar gas. The star's nature and closeness also makes it a fine probe of the local environs. From Adhara and the other stars in its neighborhood, as well as from radio and X-ray studies, we know that the Sun is in a thin coolish interstellar cloud near the edge of a hot cavity in the interstellar medium. The "hole" was created by a recent stellar explosion of the kind that may have been responsible for compressing the interstellar gases that formed the Sun and our planet.

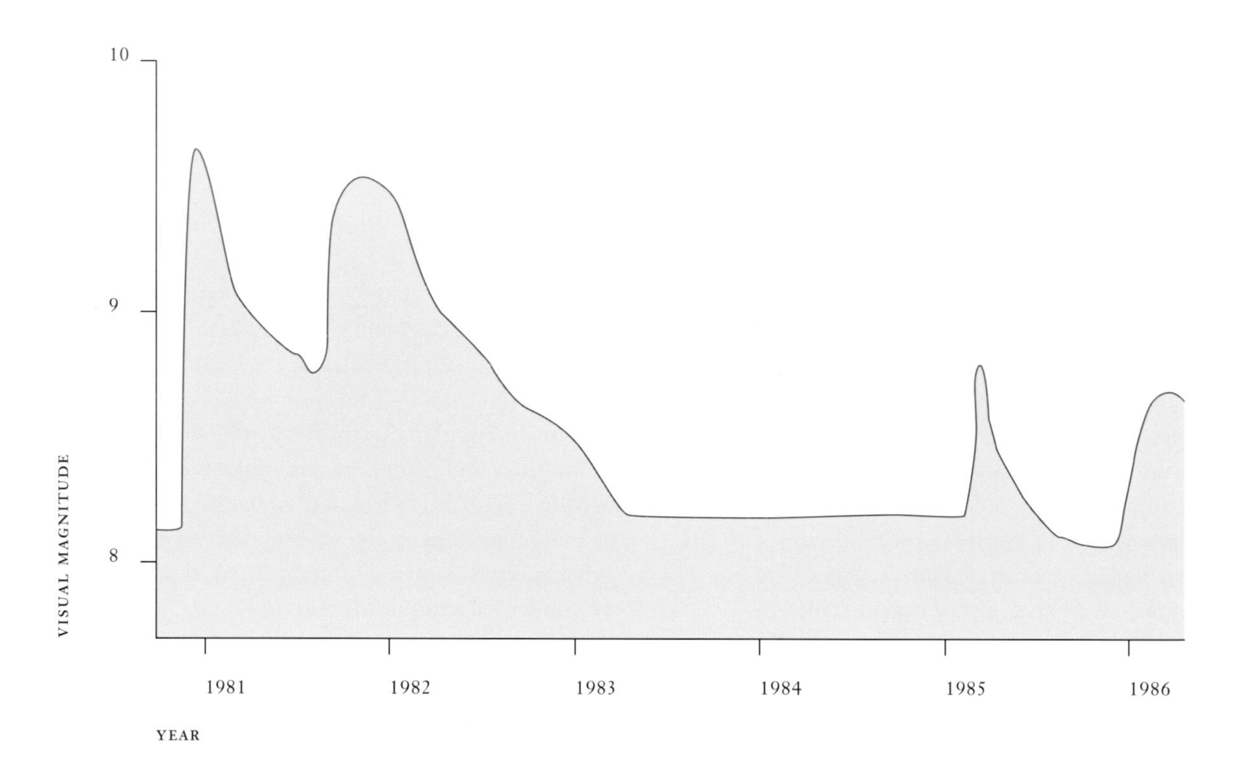

Every decade or so, AG Draconis erupts. The 1980 event went on for over five years. *(J. B. Kaler et al., 1987.* Astronomical Journal 94: 452, *and the AAVSO.)*

Residence:	DRACO
Other name:	BD+67°122
Class:	YELLOW SYMBIOTIC STAR
Visual magnitude:	9.8
Distance:	8000 LIGHT YEARS
Absolute visual magnitude:	–2.4
Significance:	AN UNUSUAL ERUPTING SYMBIOTIC.

AG DRACONIS

To break the routine of my usual observations of planetary nebulae (like **NGC 6543** and **NGC 7027**), a friend suggested I keep watch over the odd variable AG Draconis. It is so far north—67 degrees north of the celestial equator—that it is circumpolar, allowing me to observe it the year around. So every observing night for three and a half years, I recorded its brightness in 11 wavelength bands, slowly building a set of detailed light curves over nearly two and a half cycles of the star's 554-day variation period.

AG Dra is capable of powerful eruptions, but I never thought I would see one. Observing in the "old days" before targets were acquired by video, I had to preset the telescope and then locate the star through a small finder telescope. On a cold November night, with AG Dra going down in the northwest, I was leaning out over a 20-foot drop, trying to reach the finder. But the field looked wrong: one star was much too bright. After trying twice more, I realized that I was witnessing an outburst! Losing the star under the pole, I could only wait and hope the weather would hold so I could catch it coming back up in the northeast. Sure enough, there was AG Dra, a whole magnitude brighter than usual. The great 1980 eruption was in full swing.

AG Draconis is classed as a "symbiotic star," one that exhibits both low- and high-temperature features. All symbiotics are binaries that consist of a low-temperature giant star closely coupled with a white dwarf. The high-temperature features come from matter that flows from the giant to the white dwarf is being heated as it falls inward. In some cases, they may come directly or indirectly from the white dwarf itself, which may be quite hot.

Even among symbiotics—which have oddball characteristics to start with—AG Draconis is unusual. Most symbiotics contain "red giants." Classed as a "yellow symbiotic," AG Dra's giant is hotter than most. With a temperature near 4000 K, it has been placed anywhere from G7 to K4, although it is now usually taken as K0. Infrared and optical colors suggest a luminosity 1500 times that of the Sun, making this 1.5- (or so) solar-mass star a "bright giant." Lying rather far from the Galaxy's plane, at a distance of around 8000 light years, and coming at us at a great speed of 140 kilometers per second, AG Dra is clearly not a part of the Galaxy's orderly disk, but belongs to the its halo, which fits the star's low metal abundance (its iron content only 3 percent that of the Sun).

The half-solar-mass white dwarf, 1.7 AU away from the giant, is unusually hot, with a temperature in excess of 100,000 K. The white dwarf ionizes portions of the giant's wind (and even heats the white-dwarf-facing surface), creating high-excitation emission lines. Even X-rays pour vigorously from the system. Changes in the aspect of the system (yellow giant plus white dwarf plus heated wind) during the orbital period cause steady variations in the visual and ultraviolet brightness. The white dwarf also accretes matter from the giant's strong wind, which blows at a rate of one tenth of a millionth of a solar mass per year. As fresh hydrogen falls onto the white dwarf's surface, it heats. Roughly every dozen years, the temperature and density of the freshly laid surface hit a critical combination, and it erupts in a thermonuclear runaway that brightens the symbiotic binary for a period of a couple of years, typically to 5000 solar luminosities and up to 4 visual magnitudes, not unlike the behavior of a recurring nova like **RS OPHIUCHI**. So it went in 1936, 1951, 1966, 1980, and 1985, and most recently in 1994. Look again sometime around 2006.

The colorful double star Albireo (see also #86) marks the head of Cygnus the Swan. *(W. Li and A. Filippenko, University of California at Berkeley.)*

Residence:	CYGNUS
Other name:	BETA CYGNI
Class:	DOUBLE, ORANGE K3 GIANT AND BLUE-WHITE B8 DWARF
Visual magnitude:	3.08 AND 5.11
Distance:	380 LIGHT YEARS
Absolute visual magnitude:	–2.26 AND –0.20
Significance:	SHOWPIECE DOUBLE STAR WITH BEAUTIFUL COLORS.

ALBIREO

During the northern summer, Cygnus the Swan flies grandly overhead, heading south along the Milky Way as if she were fleeing the coming winter. At her tail lies the great star **DENEB**, while at her head shines a lesser light, Albireo. Though the Swan's fifth-brightest star, shining at magnitude 2.92, Albireo was designated Beta, perhaps because of its important location. The name is the result of an almost comedic mistranslation and misunderstanding, in which a scholarly reference to a word already misidentified as the star's name was itself taken as the star's name and then made to sound Arabic.

Seemingly single to the naked eye, Albireo is a well-separated double star, the pair about half a minute of arc apart. The brighter one, shining at magnitude 3.08, is a luminous but otherwise ordinary orange class K giant star, and, like **ARCTURUS**, is pausing in the process of dying, fusing helium into carbon in its core. The fainter of the two is a hot blue-white class B dwarf, a main sequence hydrogen-burner like the Sun, but just over three times as massive. As a result, the brighter, evolving star must be more massive still. Consistently, the stars are intrinsically bright, the blue one 100 times and the other 700 times the solar visual luminosity.

Ordinarily, star colors are subtle, the "blues" mere tints, the "reds" more orange, the "oranges" more yellow. But place a contrasting pair next to each other, and the eye sees the colors blossom magnificantly, the orange of the brighter star richly warm against the sharp blue of the fainter. Observers of older times grew romantic in attempts to describe the stars, Albireo's components called "golden and azure" and "topaz and sapphire."

The true duplicity of Albireo has frequently been argued. The stars are sometimes considered to fall only along the same line of sight. However, new measurements show them at nearly the same distance, the brighter K star 385 light years away, the fainter B star at 376 light years. While the difference of 9 light years—twice the distance from here to **ALPHA CENTAURI**—seems at first far too great to allow the stars to orbit each other, all observations contain inevitable and usually determinable uncertainties. Since the distances of each of the stars are uncertain by about 25 light years, they could easily be equidistant from us and in mutual orbit. However, the orbital period (over 10,000 years) would be so long that motion would be insensible.

As remarkable as the system is, each star has its own intrigue. The bright orange K star is itself double, though the two are inseparable. The combined spectrum shows that the orange star's companion is a blue class B dwarf that may be even hotter and more massive than the visually seen blue star. The close companion, however, is still no match for the orange giant, which tops them both. Albireo's dimmer blue component—the one seen through the telescope—is a "B-emission" star. Its spectrum shows that it is rotating so rapidly (over 250 kilometers per second at the equator) that it has spun off a surrounding disk. The celestial bird thus flies south with two eyes which, in turn, allow us to view a variety of stellar properties all at once.

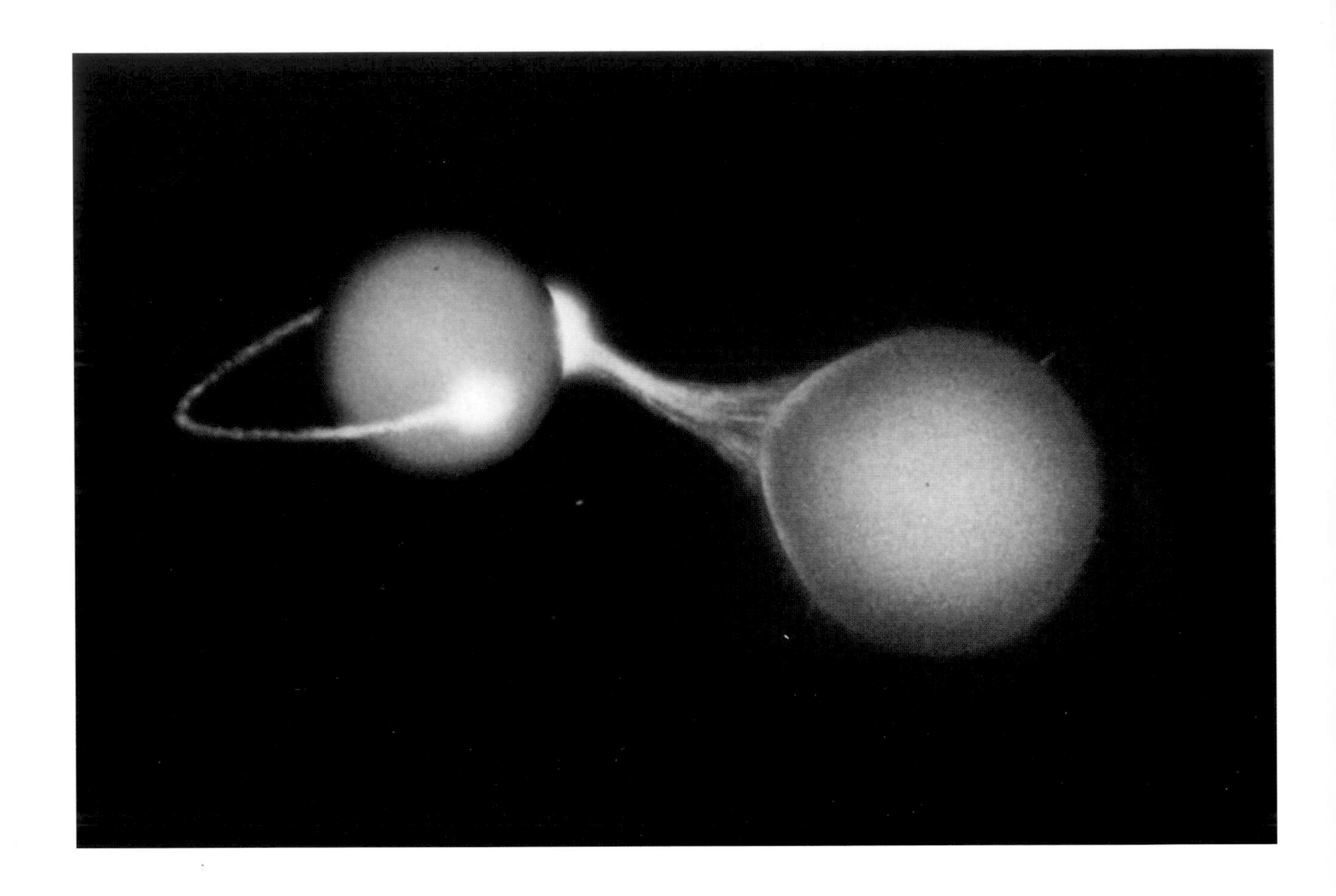

From Earth, Algol's two stars, an orange subgiant that throws matter to a blue dwarf, eclipse each other. *(S. Simpson. February 1990.* Sky and Telescope: *p.128. Based on work by D. Gillet, M. Mouchet, and P. North in* Astronomy and Astrophysics *219 [1989]: 219.)*

Residence:	PERSEUS
Other name:	BETA PERSEI
Class:	DOUBLE: BLUE-WHITE B8 DWARF AND ORANGE K SUBGIANT
Visual magnitude:	2.12 (VARIABLE)
Distance:	93 LIGHT YEARS
Absolute visual magnitude:	–0.10
Significance:	THE FIRST AND BRIGHTEST ECLIPSING BINARY STAR.

ALGOL

As the stream of stars that makes Perseus climbs the northern autumn's sky, look south toward Andromeda to Algol, which contains a delightful surprise. Every 2.867 days, the star dims for a few hours, dropping to 30 percent its original brightness. Algol, the "Demon Star," is a prime example of the eclipsing binary in which two stars are locked in tight gravitational embrace, their orbital planes lying nearly in the line of sight. As the pair of stars revolve, each gets in the way of the other, leading to a pair of eclipses.

While thousands of eclipsers are known, Algol represents a special class that consists of a small bright star and a much dimmer, but larger one. In Algol's case, a hot blue-white class B main sequence dwarf mutually orbits a class K "subgiant," a star that has ceased core hydrogen fusion, has a quiet helium core, and is trying—unsuccessfully—to expand to larger proportions. The B star, with a temperature around 12,000 K, is visually some 100 times more luminous than the Sun and triple its size. The other star has a temperature of about 4000 K, is visually only 4 percent as bright as its mate, and just a little larger. The two are separated by about 0.07 AU, 20 percent of the distance of Mercury from the Sun. The main eclipse occurs when the brighter star is partially hidden by the larger, dimmer one. Midway between primary eclipses, the smaller star passes in front of the larger, and the light barely dips for a few hours—an event unnoticeable to the eye.

Using data collected during an eclipse, astronomers can discover a great deal about the eclipsers themselves, making these binaries vital to learning about stars in general. From the graph of magnitude plotted against time, we can find the tilt of the orbit. Spectroscopic data then give the orbital velocities, from which the nature of the orbit can be found, leading to stellar masses. From the durations of the eclipses, we can even tell stellar dimensions. Mass measurement in the Algol system is special in that it reveals a curious paradox. Higher-mass stars have shorter lives because their fuel is used much more quickly. The mass of the bright B star is three to four times solar. The dimmer dying companion, however, is contrarily the less massive of the two, only about 60 percent the solar mass. The only explanation is that the dim companion was indeed once the more massive, but has lost much of its matter.

All bodies in close proximity raise tides in each other, but here, the tides occur in grand proportion. Surrounding any member of a double star is an invisible "tidal wall," where, as a result of the combined gravity and the orbital motions of the stars, the escape velocity is effectively zero. As Algol's dimmer star aged and expanded, it ran into the wall, where its mass piled up and began to overflow to the hotter and more compact star. The process eventually reversed the original mass ratio. The effects of the gas stream, which circles and crashes into the class B star, are readily visible. And the effects will continue. Most stars of modest initial mass end up as white dwarfs made of carbon and oxygen that are simply old, burned-out stellar cores. The dim star of Algol, however, will probably never make it that far because its internal helium will be unable to become hot and dense enough to fuse to carbon, and it will probably die as a rare helium white dwarf. Go watch the show for yourself.

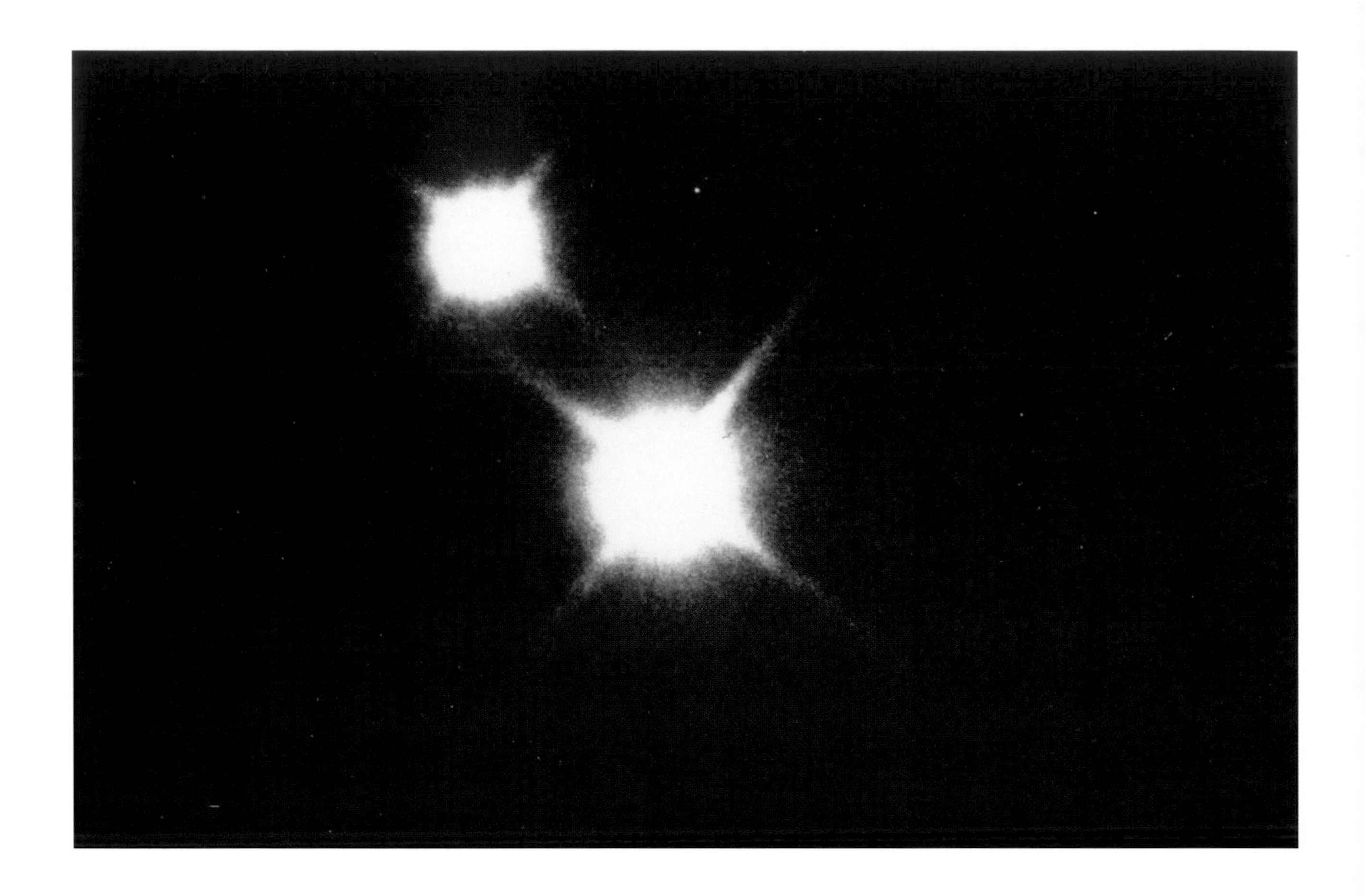

The dimmer class K component of Alpha Centauri (see also #1) and the brighter class G star orbit each other every 80 years. *(NOAO/AURA/NSF)*

Residence:	CENTAURUS
Other name:	RIGIL KENTAURUS (AND PROXIMA)
Class:	G2 + K1 + M5 DWARFS
Visual magnitude:	–0.01, 1.33, 11.05
Distance:	4.22 TO 4.39 LIGHT YEARS
Absolute visual magnitude:	4.34, 5.68, 15.50
Significance:	THE CLOSEST STELLAR SYSTEM.

ALPHA CENTAURI

With some exceptions, astronomers must look at what Nature has to offer. The result is a recurring problem of "observational selection." The apparent sky is dominated by relatively rare but very luminous stars that can be seen over great distances. Of the 50 brightest stars, all but one are at least seven times more luminous than the Sun, and most are hundreds to thousands of times more luminous. The exception is Alpha Centauri—Rigel Kentaurus, the "Centaur's Foot." Ranking third behind **SIRIUS** and **CANOPUS**, it appears bright only because, at four light years, it is the nearest of all stars.

Naked-eye Alpha Centauri, however, is not one star but two. The pair is easily resolved with a small telescope. The duo orbit each other every 79.9 years separated by an average of 24 AU, somewhat greater than the distance between Uranus and the Sun. Both are Sun-like dwarfs with apparent magnitudes of –0.01 and 1.33. If we could see them separately with the naked eye, the brighter would still be the third-brightest star, while the fainter would rank a very respectable 21st. This secondary class K component is cooler than the Sun, its temperature about 5000 K. With a mass of 0.85 solar, its luminosity is only about 40 percent solar. The brighter class G2 component is a near-solar clone with a mass only about 10 percent greater than the Sun's. However, we do not know of any planets in orbit around either of the stars, or how well a planet would survive such a double-star system. But assuming there was a populated planet, its people would look back on us and see our Sun as the eighth-brightest in their sky, shining at magnitude 0.48 midway between the stars of Cassiopeia and Perseus.

A bit over 2 degrees to the southwest and 10,000 AU closer lies the third member of the system, dim Proxima Centauri. Orbiting Alpha proper with a period of 1 million years or more, no relative movement can be detected. We think it belongs to the system only because it and its apparent mates move together through space with about the same direction and speed. Though a mere 4.2 light years away and among the brighter M dwarfs (none of which can be seen with the naked eye), it is still only an 11th magnitude star, over 60 times fainter than the eye can see, radiating 1/18,000 of the visual light of the Sun. To be visible to the naked eye, the star would have to be eight times closer to us, a distance about 800 times the distance from here to Pluto. Even from the bright G and K double, Proxima would shine only at fifth magnitude, as bright as the faintest star of the Little Dipper. The star's luminous parsimony is a result of its low mass, about two-tenths solar. On the plus side, the nuclear reactions that support Proxima run so slowly that it will live for 1 trillion years, far outlasting its bright companions from which it may well escape as a result of Galactic tides (if it has not already done so).

To receive the same amount of energy from Proxima that we do from our Sun (factoring in the infrared radiation from the cool 3500-K surface), we would have to orbit 6 million kilometers away with a period of three Earth days. Life, however, would be impossible. The rotating Sun produces magnetic fields that upon collapse can generate powerful X-ray "flares" as big as the Earth. Those on Proxima can involve the whole star! Proxima and great numbers of similar "flare stars" erupt visually by one magnitude or more and also produce intense X-rays. Weather forecast for today: "warm and sunny—and, oh, the sun might suddenly become three times as bright as normal for a few minutes, so wear a hat."

The barium in barite, a mineral made of crystals of barium sulfate, was created almost entirely in evolving giant stars like the now-dead companion to Alphard. *(J. B. Kaler.)*

Residence:	HYDRA
Other name:	ALPHA HYDRAE
Class:	ORANGE K3 GIANT
Visual magnitude:	1.98
Distance:	175 LIGHT YEARS
Absolute visual magnitude:	–1.69
Significance:	ALPHARD: THE "LONELY ONE," AND ALMOST THE BRIGHTEST BARIUM STAR.

ALPHARD

Though stars are scattered across the sky more or less at random, we find both crowds and near-empty voids. Within the voids, bright stars stand out. And so it is with Alphard, the brightest star in Hydra, "the Water Serpent," which slithers a third of the way around the sky. At first appearance, Alphard is just another orange class K giant rather like **ARCTURUS**, though more luminous. Mid-second magnitude even though 175 light years away, it is visually 400 times more radiant than the Sun. When the invisible infrared light radiated by the coolish 4000-K surface is factored in, the luminosity climbs to 750 times solar. To achieve such brilliance requires great size, the star almost 60 times bigger than the Sun. Placed in the Sun's position, Alphard would extend 70 percent of the way to the orbit of Mercury.

The star would be of little interest were it not for its peculiar spectrum. Alphard is a bright, though mild, example of a "barium star." Such stars have abnormally strong blue-violet absorptions caused by ionized barium, number 56 in the periodic table. Along with barium, we also see enhancements of carbon, strontium, and other elements.

A triumph of twentieth-century astrophysics was the discovery that all the chemical elements except for hydrogen, helium, and a little lithium are made in stars. Our Earth is constructed from the leavings of earlier stellar generations. Barium is created deep within giant stars that have dead, burned-out carbon cores, which are surrounded by shells where helium is fusing to yet more carbon. Within such a helium-burning shell, an atom heavier than iron can capture free neutrons one at a time, which increases the atom's isotope number. When enough neutrons have been caught, the atom becomes radioactive. At that point, a neutron in the nucleus will change to a proton by spitting out a negative electron, which causes the original atom to become the next-heavier chemical element. Through this "slow neutron capture," cesium becomes barium, barium becomes lanthanum, and so on. These freshly created elements can subsequently be brought to the star's surface by convection.

However, Alphard and its cohort of barium stars have the wrong structures to make the stuff themselves. Instead of helium-burning shells, they have helium-burning cores. The structures required for slow neutron capture are reserved for more advanced stars like **MIRA**. So where does the element come from? Diligent analysis of the barium stars shows them to be doubles with shrunken white dwarf companions like **SIRIUS B**, stars that at one time were the nuclear burning hearts of normal stars like the Sun. To become a white dwarf, a star must go through its entire life cycle, including through the phase in which such elements as carbon, barium, and others can be dredged up. While an advanced giant, the now-white dwarf must have passed matter onto its then-ordinary dwarf companion (much as the evolved star of **ALGOL** is doing now), contaminating it with heavy elements. The star we now see as Alphard began its own death process after its companion had already become a white dwarf, and we now see the contaminants loaded into Alphard proper. This process is one more result of stellar interchange that can lead not just to contamination but, as seen elsewhere, to violence beyond imagination.

A glowing dusty nebula around reddish Antares (see also #79), caused by aeons of mass loss, reflects the red supergiant's light. *(© Anglo-Australian Observatory/Royal Observatory, Edinburgh, photograph from UK Schmidt plates by David Malin.)*

Residence:	SCORPIUS
Other name:	ALPHA SCORPII
Class:	M1 RED SUPERGIANT
Visual magnitude:	0.96
Distance:	620 LIGHT YEARS
Absolute visual magnitude:	–5.40
Significance:	FIRST MAGNITUDE RED SUPERGIANT NEAR THE ECLIPTIC.

ANTARES

Scorpius: there is no wondering why the constellation was so named, as it looks for all the world like the feared arachnid, its curved deadly tail swooping downward, with the great red supergiant star Antares at its heart. Antares is one of Nature's practical jokes. The star is about the same color and apparent brightness as Mars, which also plies the Zodiac, confusing beginners and giving Antares its name, "*Ant Ares*," from Greek "like Ares," Ares the Greek version of the god of war.

Antares is one of only two first magnitude supergiants. The other is Orion's **BETELGEUSE**. Both are magnificent stars that dwarf the sky's more ordinary giants like **ARCTURUS**. Cleverly placed opposite Betelgeuse, one or the other is almost always visible. From Antares's measured angular diameter and a rather uncertain distance of 620 light years, we calculate a radius over 800 times that of the Sun, almost 4 AU, which is about 70 percent the size of the orbit of Jupiter! Visually 12,000 times brighter than the Sun, its cool 3400-K surface radiates powerfully in the invisible infrared, which, when taken into account, boosts the luminosity to 90,000 or so solar luminosities (although uncertainty in the distance leaves considerable room for error). The star's redness shows that it is highly evolved. Having started as a heavy, blue main sequence star of 15 to 20 solar masses, its core is most likely fusing helium to carbon. The required readjustments have expanded the outer envelope to its current immense proportions and have made the star somewhat unstable, causing to vary slightly in brightness.

Unlike most constellations, Scorpius is not just a collection of random stars. Most are hot and blue, and are part of a large, disintegrating "association" of massive stars. All are at about the same distance and were born more or less together from a vast parent interstellar cloud. In that respect, this grouping is very much like Orion, which contains the other red supergiant. Antares is among the most massive stars of its association and has evolved first to grace the others with its pale ruby light.

Antares's huge size and luminosity promote a fierce wind that blows at a rate of nearly 1 millionth of a solar mass per year. The star has encompassed itself in its own expanding effluvia, some of which has condensed into dust that is illuminated by reflected Antarian starlight. Buried within this cloud is a much smaller, thick dust shell that reveals recent activity. Antares and its nebula also harbor a surprise, a hot, blue sixth magnitude class B companion three seconds of arc (450 AU) away that has not yet begun to evolve and that, by contrast, looks almost green through the telescope. The companion, lying deep within the dusty ejecta, has carved an ionized cavity within the great wind about 1000 AU across. It is useful because its simple spectrum provides a fine background against which to study the outflowing wind.

Enjoy Antares now, because it probably does not have much time left to it. Sometime in the next few million years, its internal fusion will advance from helium burning to carbon-burning and onward toward the creation of an iron core that will collapse, sending the star into a catastrophic explosion. The resulting supernova would be visible not only across the Milky Way but to anyone in other galaxies looking back at us. And the scorpion would finally be crushed beneath an iron heel.

The Yerkes Observatory 40-inch telescope, the world's largest refracting telescope (completed in 1897), used Arcturus's (see also #56) orange light to open the 1933 Chicago World Fair. *(Courtesy of Yerkes Observatory, University of Chicago.)*

Residence:	BOÖTES
Other name:	ALPHA BOÖTIS
Class:	ORANGE K1 GIANT
Visual magnitude:	–0.04
Distance:	37 LIGHT YEARS
Absolute visual magnitude:	–0.30
Significance:	THE BRIGHTEST STAR IN THE NORTHERN HEMISPHERE; ANCIENT LINEAGE.

ARCTURUS

Among the very brightest of stars, Arcturus illuminates spring and summer skies with a soft yellow-orange light. With **VEGA** and **CAPELLA**, it composes a trio of luminaries that partition the northern sky. Of the three, Arcturus is slightly the brightest, making it the most radiant star of the Northern Hemisphere and the fourth-brightest star on the celestial sphere.

Arcturus, the "Bear Watcher," follows Ursa Major, the Great Bear, around the northern pole. The star's name derives from "*arktos*," the Greek word for "bear," and our word "arctic" references the Bear's northerly position. Yet the star is far enough south to be seen by everyone north of Antarctica. Not uncommon among bright stars, Arcturus has an ancient lineage and is among the first stars mentioned by name. Steeped in lore, Boötes figures in the Odyssey, where the name refers not to the constellation but to the brilliant star. Nearly 3000 years later, the star is no less famous. In the early twentieth century, Arcturus's distance was placed at 40 light years. Since 40 years was the interval between the two great Chicago World Fairs of 1893 and 1933, the light that had left the star during the first Fair would have arrived just in time for the next. Arcturus seemed the perfect celebrity to be asked to open the second Fair, and a photocell at the focus of the University of Chicago's Yerkes Observatory 40-inch telescope activated by the star began the festivities by turning on the Fair's lights.

A classic orange giant with a surface temperature of 4300 K, Arcturus shines 215 times more brightly than our Sun and has a diameter nearly 25 times solar. If it were to replace the Sun, the star would appear 12 degrees across in our sky, more than double the separation between the front bowl stars of the Big Dipper. It is both close and large enough for a highly accurate measure of its angular diameter: 25 thousandths of a second of arc. Like all giants, Arcturus is a dying star and has probably begun to fuse its core helium into carbon and oxygen, dimming and heating some as it adjusts to a long period of quiet helium-burning. About twice as massive as the Sun, Arcturus is now about 1 billion years old, some 20 percent the age of our Solar System, and its larger mass is causing it to burn out more quickly. When the core helium runs out, Arcturus will expand well beyond its present confines, lose its outer shell, and produce a planetary nebula. It will finally quit stellar life as a dim, dense white dwarf.

Having a higher velocity than other bright stars, Arcturus does not quite share the same orbital characteristics as the Sun. It comes from a somewhat older population of the Galaxy, one whose stellar orbits are not in the plane of the thinnest part of the Galactic disk. Since the metal content of our Galaxy increases with time as processed gases are ejected by dying stars, the older parts of the Galaxy generally have lower metal contents. Though the correlations between Galactic orbit and heavy-element content are not always very clear, Arcturus is consistently somewhat deficient in metals, only having about 20 percent as much iron relative to hydrogen as found in the Sun.

Watch for this flower of the northern spring sky as its earthly counterparts begin to bloom.

10

Barnard's Star scoots quickly across the sky, as seen in these pictures made in 1937 (top) and 1960 (bottom). *(Courtesy of Lowell Observatory.)*

Residence:	OPHIUCHUS
Other name:	BD+04°3561
Class:	M5 RED DWARF
Visual magnitude:	9.54
Distance:	6 LIGHT YEARS
Absolute visual magnitude:	13.30
Significance:	THE FASTEST-MOVING STAR, FAMED FOR THE PLANETS THAT WERE NOT THERE.

BARNARD'S STAR

As the stars go about the Galaxy on slightly different orbits, they change their positions relative to each other. As a result, the constellations we see every night are temporary. The Dippers will eventually begin to leak, Orion will depart for warmer climates, and Cancer might actually begin to look like something. A time-traveller venturing several million years into the future would find a completely unfamiliar night sky.

For us, however, such motions are exceedingly slow and unnoted, as the constellations change over a time longer than the history of the printed word. Dim Barnard's Star is an important exception. Discovered by E. E. Barnard at Chicago's Yerkes Observatory in 1916, Barnard's Star veritably screams across the sky at a break-neck pace of 10 seconds of arc per year. While this displacement is but the angular size of a US nickel seen at a distance of half a kilometer, it is nevertheless a celestial record. Over the average human lifetime, the star will move by two-tenths of a degree, nearly half the angular diameter of the full moon. Proceeding almost due north, Barnard's Star will leave Ophiuchus for Hercules (on its way to Draco) in about 3300 years. At its distance of six light years, this rate of motion translates into a speed of 90 kilometers per second across the line of sight. Combined with its velocity of 108 kilometers per second along the line of sight, the speedster zips through space at 140 kilometers per second, a rate which is seven times that of the average star.

Not that all this hurrying matters much to the casual viewer, because Barnard's Star is 15 times fainter than the unaided human eye can see. Other than its great speed, it is an ordinary hydrogen-fusing, cool (about 3500 K) class M red dwarf, just one of a huge number that flock around the Sun. Over 70 percent of all stars fall into this and cooler varieties, but they are so faint that none is visible to the eye alone. It would take 2400 "Barnard's Stars" to equal the apparent luminosity of our Sun. This dimness is offset somewhat by the fact that cool Barnard's Star radiates much of its light in the invisible infrared. If that is taken into account, the little star (only 0.2 times the solar radius) would appear 11 times brighter. If the star were to replace our Sun, it would appear as a tiny reddish disk dimly illuminating an ice-covered Earth on which we would receive about as much energy as a body orbiting halfway between Saturn and Uranus!

The origin of Barnard's Star's faintness lies in its low mass, which is only about one tenth that of the Sun. As a result, what little fuel it has "burns" feebly. But what it lacks in brilliance is made up for in longevity. Long after the Sun has departed for the land of the white dwarfs, Barnard's Star will still be with us, though far out of sight in some other part of the Galaxy.

Its fame very much out of proportion to its actual brilliance, Barnard's Star centered a great controversy in the 1960s when astronomers thought they noted it wobbling back and forth on its inexorable path against the Ophiuchan stars. The small motion was attributed to the gravitational effects of Jupiter-like planets. Alas, the "wobble" was apparently the result of changes in the position of the telescope's lens. Barnard's Star is now officially "planetless," though that status could dramatically change as technology improves.

The oscillations of a giant floating soap bubble rather resemble the multiple pulsations of Beta Canis Majoris (see also #2) stars. *(J. B. Kaler.)*

Residence:	CANIS MAJOR
Other name:	MIRZAM
Class:	BLUE CLASS B1 GIANT, OR BRIGHT GIANT
Visual magnitude:	1.98
Distance:	500 LIGHT YEARS
Absolute visual magnitude:	3.94
Significance:	CLASSIC MULTIPLE-PERIOD VARIABLE STAR.

BETA CANIS MAJORIS

SIRIUS, the brilliant eye of the Hunting Dog who follows Orion, is so important that two stars herald its coming. Procyon in Canis Minor (the smaller Hunting Dog), whose name means "before the dog," rises first, and then comes our star, Beta Canis Majoris, which is more rarely called Mirzam.

Like so many stars of the northern winter, Mirzam is a hot, blue class B star of the kind that gives so much wondrous sparkle to a sky seen above a snow-covered countryside. About 500 light years away, it shines with a luminosity (including the ultraviolet light from its very hot 22,000-K surface) 19,000 times that of the Sun. A truly great star ten times larger than the solar diameter, our Earth would have to orbit at a distance of 140 AU for us to survive—over thrice the distance of Pluto from the Sun.

Mirzam, born with a mass somewhere around ten times that of the Sun, is dying. Now in the process of adjusting to a changing interior with a dead helium core, the vast hydrogen envelope will slowly expand and cool, ultimately producing a red giant or supergiant. Like **ADHARA**, Mirzam has a simple spectrum and generates enough light to make a fine background star with which to probe interstellar space. Both stars lie along an "interstellar tunnel," a rarefied void in which there is little or no neutral gas.

Mirzam's claim to stellar prominence lies not in these properties, but in its variability. A great many stars change in brightness, not just because of stellar eclipses like that produced by **ALGOL**, but as a result of intrinsic instabilities that cause the stars to pulsate, which, in turn, continuously changes the stars' temperatures and luminosities. Some are anything but subtle. **MIRA**, the prototype for the "long-period variables," changes in the optical spectrum by over seven magnitudes during the course of about one year. **DELTA CEPHEI**, the archetype of the famed "Cepheids" and a better comparison, changes by about one magnitude over a very regular period of 5.37 days.

Mirzam varies by what would seem to be an unimportant tenth or so of a magnitude. Unlike the simple variations seen among the "Cepheids" and "Miras," Mirzam's are nothing short of weird. The star is one of a class epitomized by fainter Beta Cephei, the first of the kind discovered. As a result, this set of stars is confusingly known as both "Beta Cephei" and "Beta Canis Majoris" variables. Such stars vary quickly and not with one period, but with several periods at the same time. Mirzam, for example, oscillates with periods of 0.2500, 0.2513, and 0.2390 days, as well as with other periods. The different periods beat against each other like out-of-tune guitar strings, resulting in a seemingly erratic variation. Moreover, the pulsations do not affect the star's radius in a simple way. Instead of the whole star alternately becoming larger or smaller, some parts of Mirzam and its cohorts pulse outward and others simultaneously move inward as the star oscillates rather like a thrown water balloon.

Beta Cephei stars, all hot class B giants, are quite similar to each other. Why they vary in this way is uncertain. And why stars that are almost the same, like Adhara, do not so pulsate is anyone's guess. Apparently, all stars of around 10 to 15 solar masses go through a similarly erratic phase as they begin to evolve, temporarily developing very specific internal structures. We do not know why.

12

The celestial sphere encloses Earth in this antique representation of the sky. The equinoctial colure, marked by Beta Cassiopeiae (See also #43 and #95), runs across the middle, while the solstitial colure (defined by the Sun's solstices) circles the periphery. *(Andreas Cellarius. 1708.* Harmonica Macrocosmica seu Atlas Cosmographiam. *Courtesy of the Rare Book Room and Special Collections Library, University of Illinois.)*

Residence:	CASSIOPEIA
Other name:	CAPH
Class:	F2 GIANT OR SUBGIANT
Visual magnitude:	2.27
Distance:	54 LIGHT YEARS
Absolute visual magnitude:	1.16
Significance:	A TRIPLE-THREAT "COLURE" STAR, RARE "GAP" STAR, AND THE BRIGHTEST "DELTA SCUTI" STAR.

BETA CASSIOPEIAE

High overhead, the "W" of Cassiopeia looks like outstretched fingertips. Indeed, the Arabs called the collection "the stained hand" (a reference to cosmetic coloring), from which the Beta star received the shortened Arabic name "Caph." Barely fainter than the Alpha star, Beta nicely tops its neighbor, **GAMMA CASSIOPEIAE**, and certainly equals it in importance. It is another of the "marker stars" that, like **POLARIS**, reveal the sky's points and circles. As the Earth orbits, the Sun appears to traverse an "ecliptic" path that is tilted to the celestial equator (marked by **DELTA ORIONIS**) by 23.4 degrees. As a result, the Sun moves annually from north to south and back, giving us the seasons. On March 20, the Sun crosses the celestial equator from south to north at the vernal equinox in Pisces, beginning northern spring. On September 23, it again travels back to the south at the autumnal equinox in Virgo. The circle that embraces the equinoxes and the celestial poles, running perpendicular to the celestial equator, has a marvelous mouthful of a name, the "equinoctial colure." If you draw a line southward from Caph to Alpheratz (Alpha Andromedae), it will parallel the colure just 2 degrees to the west and point close to the vernal equinox.

From its distance of only 54 light years, this class F2 giant radiates 28 times more light than the Sun from a 6700-K surface. The star is only four times the solar size, not much to say for a "giant," and that is what makes it special. Sometimes classed as a "subgiant," Caph is intent on becoming a real giant. Stars begin to die when the abundant hydrogen that fuels their cores runs out, and the now dead-helium cores begin to contract and heat. Over time, the outer hydrogen envelope expands and cools, reddening the star to class M. When the interior becomes hot enough, helium fires up to fuse to carbon, and the star stabilizes as a warmer (G or K) giant.

The transition to gianthood, in which a modest-mass star is cooling through classes F and G (the "Hertzsprung Gap," after the Danish astronomer Ejnar Hertzsprung) is very fast, so such stars are rare. Caph, with its contracting core, will spend only about 1 percent of its life in "the gap," where it epitomizes the beginning of stellar death. Its state is similar to the fainter component of **CAPELLA**, but Caph can be seen by itself in all its odd glory. Even within this group, it is rare, as most of these stars have notable coronae, or surrounding magnetically heated hot atmospheres. Caph's is oddly and particularly weak.

The internal structures of mid-range supergiants of classes F and G become unstable. The luminious supergiant, **DELTA CEPHEI**, for example, dramatically halves its brightness every 5.4 days, the variations and periods of such "Cepheids" increasing with mass and luminosity. Though the effect greatly diminishes down near the luminosities of the dwarfs, stars in classes A and F still continue to chatter away in a similar but more complex and subdued manner. Caph is the brightest of these "Delta Scuti" stars, varying by only 6 percent over a period of 2.5 hours.

A star in transition, Caph was once a two-or-so solar mass class A main sequence star, not unlike **VEGA** is today. When it becomes a red giant, its Delta Scuti character will be long forgotten. Even its place as a colure star is being lost. While precession is making both Polaris and Delta Orionis better markers, Caph is becoming worse, having crossed the colure in 1828. But have heart. Though Beta Cassiopeiae will never again be a main sequence star, it will pass through the other side of the colure in a mere 12,800 years.

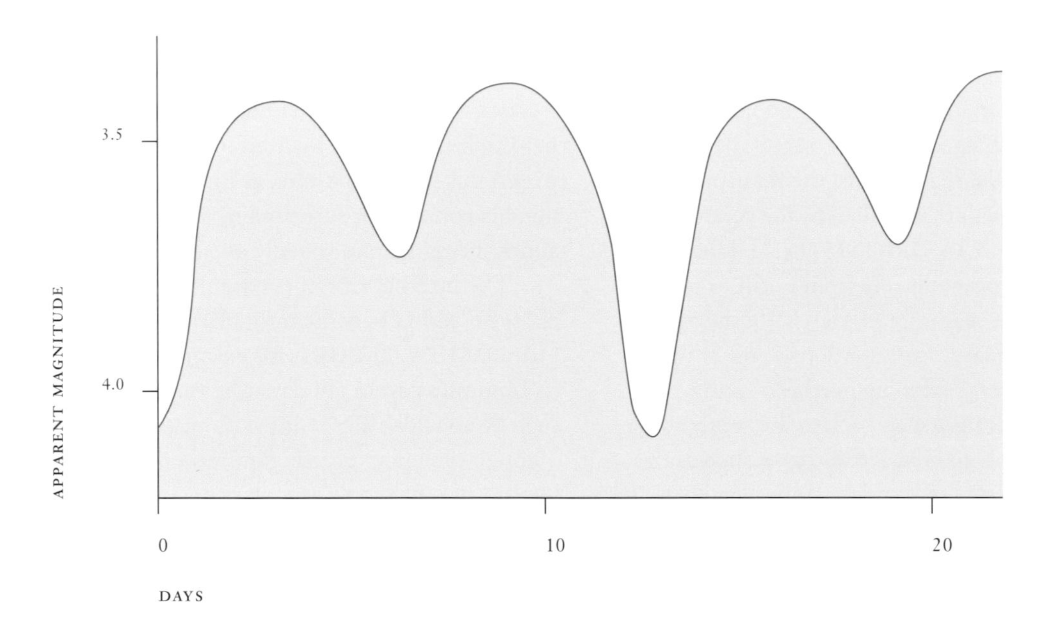

The continuously-changing light of Beta Lyrae (see also #35), just south of VEGA and EPSILON LYRAE, is the result of highly distorted, mass-exchanging stars. *(Adapted from R. Burnham, Jr. 1978.* Burnham's Celestial Handbook. *New York: Dover.)*

Residence:	LYRA
Other name:	SHELIAK
Class:	BLUE B7 BRIGHT GIANT AND SOMETHING ELSE
Visual magnitude:	3.45 (VARIABLE)
Distance:	900 LIGHT YEARS
Absolute visual magnitude:	–3.71
Significance:	A KEY MASS-TRANSFER DOUBLE AND ONE OF THE LEAST UNDERSTOOD OF THE BREED.

BETA LYRAE

A little parallelogram drops from Lyra's brilliant **VEGA** to form the body of the Celestial Harp, with third magnitude Beta and Gamma Lyrae at its southern end. Using Gamma Lyrae as a reference, you can see that Beta changes in brightness, dropping from its maximum of 3.45 down to mid-fourth magnitude near 4.0 and back again over a period of 12.94 days. Though Vega is more famed, some would say that Beta Lyrae (whose alternative name "Sheliak" comes from the Arabic word for "harp") is by far the more important.

From its distance of 880 light years, the visual luminosity is 2600 times that of the Sun. Beta Lyrae's spectrum and blue color clearly indicate the presence of B stars. The nature of the light curve and the precision of the period of variation (measured to within 1 second) tell us that the star is an eclipsing double rather like **ALGOL**. Beyond these simple deductions, wresting information from the system has been monumentally difficult. The star is one of the most studied of all time, over 500 research papers referring to it in the second half of the twentieth century alone.

Sheliak's optical spectrum reveals the light of just one star, a blue class B7 (or so) bright giant. The second component, separated from its mate by just 30 solar diameters, is the more massive and seems to be a class B dwarf that is almost hot enough to be an O star. But it is too faint and must somehow be obscured. The spectrum also shows a wealth of emissions from highly ionized atoms, indicating hot, free-flowing gas. The resulting picture, developed long ago, is that of a close pair of hot and massive B stars—one evolved, the other not. The giant has hit the "tidal wall" and is losing matter to the dwarf. Unlike Algol, however, the matter does not flow directly to the gaining star, but first into a thick disk that surrounds it, one so fat and dense that the B dwarf is almost completely hidden. Moreover, about half the matter does not make it to the dwarf at all. Instead, it flies free into interstellar space, some of it in twin flows that run perpendicular to the plane of the orbit. The result of the interchange and lost mass is that the orbital period lengthens at the majestic pace of 9 seconds per year: stellar evolution in a human lifetime.

This fine mess is very difficult to analyze. Because of tidal distortions and radiating circumstellar matter, Beta Lyrae varies continuously, even when not in obvious eclipse, making it hard even to find out when the eclipses begin and end. The visible giant has a surface with a temperature of 13,000 K radiating over 6000 solar luminosities. In its death throes, the star has expanded to 15 times the solar diameter. The obscured mass-gainer is much grander, a 30,000-K star radiating 25,000 times the solar brightness compacted into a body less than half the size of its companion. The giant, though visibly brighter, is actually the fainter as a result of the stuff that hides the accreting star.

The evolved star must have begun life as the more massive of the two. Yet the other star, the gainer, along with its disk now outweighs it by a factor of four. The "loser" is now far less than a third of its original mass, while the gainer is about 50 percent heavier! Beta Lyrae seems to be nearing the end of its most vigorous mass-tossing phase and, although it is still a mystery in many ways, gives us a great insight into how close doubles evolve.

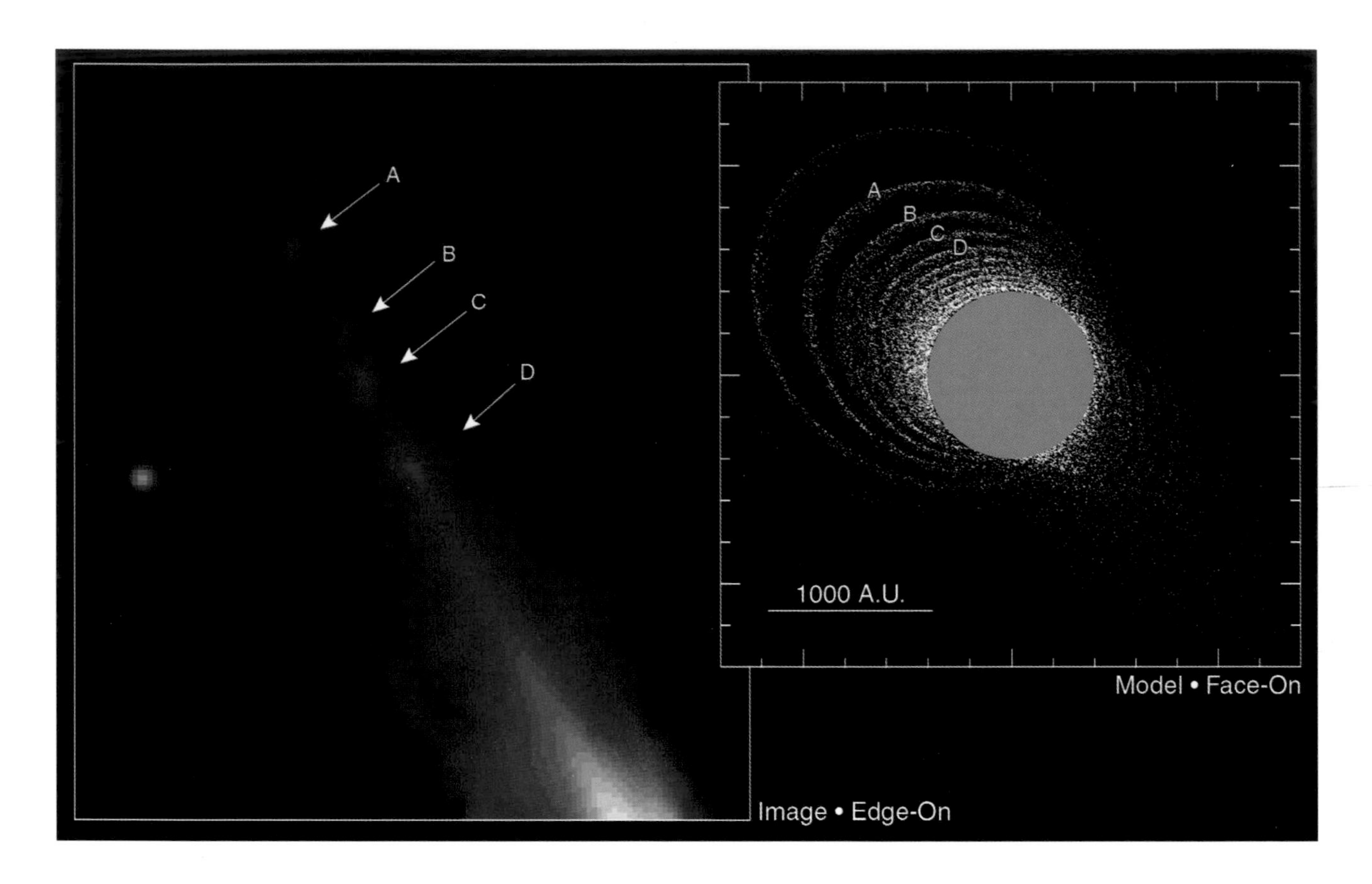

A thin dusty disk extends from Beta Pictoris. Embedded elliptical rings may have been made by a passing star. *(Hubble Space Telescope, P. Kalas [StScI], J. Larwood [Queen Mary and Westfield College], B. Smith [University of Hawaii], A. Schultz [STScI], and NASA.)*

Residence:	PICTOR
Other name:	HR 2020
Class:	ORDINARY WHITE CLASS A5 DWARF
Visual magnitude:	3.85
Distance:	63 LIGHT YEARS
Absolute visual magnitude:	2.42
Significance:	THE FIRST MATURE STAR SEEN TO HAVE WHAT APPEARS TO BE A SURROUNDING PLANETARY DISK.

BETA PICTORIS

Little did Nicolas de Lacaille know that his obscure southern constellation Pictor (the "Painter's Easel"), invented in the eighteenth century to fill a space between bright constellations, would become famous for holding one of the sky's most intriguing stars. Outwardly it is just one more warm (8300 K) class A dwarf, one much like **SIRIUS** or **VEGA**. Nearby, at a distance of only 19.2 light years, this two-solar-mass star shines with a luminosity nine times that of the Sun.

Though seemingly ordinary, Beta Pictoris is a vivid illustrator of the wonders that can be achieved with sophisticated instrumentation. The key to the star's fame lies not in what we can see with the eye, but in the longer-wave infrared. First discovered by William Herschel in 1804, the infrared wavelength band languished until the 1960s as a result of poor detectors. By 1982, astronomers had developed superb detectors that orbited the Earth aboard the Infrared Astronomical Satellite (IRAS), which mapped the sky in four wavelength bands. Among IRAS's discoveries were a handful of normal stars that were surrounded by clouds of matter, and that the infrared radiation was coming from embedded cool dust. Our Solar System formed from a disk of dust that congealed to make larger bodies. Could we be seeing the remnants of planetary formation?

At optical wavelengths, the clouds should be illuminated by reflected starlight and are devilishly difficult to see because of the brightness of the embedded stars. Ground-based observations of Beta Pictoris, which employed a device that blocked out direct starlight, showed that the cloud was in the shape of an edge-on disk, strengthening the speculation that we are either seeing a planetary system in the making or one that might already have been completed.

Additional observations over the years have only increased that likelihood. The disk is some 400 AU in radius—about ten times the diameter of our planetary system out to Pluto. However, Pluto is not the end of the Solar System. Astronomers are certain that a vast belt of debris spreads outward in a thick disk far beyond Pluto. Called the Kuiper Belt, it is the source of short-period comets, the matter too thinly spread to have made a planet. Over 600 individual bodies have been found there, and Pluto is now considered to be the biggest of them. Moreover, the Solar System is filled with smaller particles that result from collisions among larger bodies, and the Kuiper Belt is almost certainly filled with some of that dust.

Perhaps the Beta Pictoris disk is the star's "Kuiper Belt," though one much thicker than our own. Infrared spectroscopy reveals that the dust is made of carbon and silicates, the same kind of stuff we find in interstellar space and which is thought to have been present in our own forming planetary system. More sophisticated observations with the Hubble Space Telescope hint at blobs of matter falling toward the star. Are they comets of some sort? Hubble and ground-based observations show a central "hole" in the disk where the dust is diminished, suggesting that planets have indeed formed from the dust, as happened in our own Solar System. Moreover, the disk is warped, perhaps as a result of the gravity of one or more large planets.

All the current evidence is indirect. No one has seen a planet in the Beta Pictoris disk. For that matter, no one has actually seen a planet orbiting any star. Our strongest evidence for extrasolar planets is in the reflexive motions seen in stars like **51 PEGASI**. Eventually we will see one, and it would be no surprise if Beta Pictoris did not lead the way.

Betelgeuse (see also #27), one of few stars whose surface has been directly imaged, displays an unexplained hot spot. *(Hubble Space Telescope, A. Dupree [Harvard-Smithsonian Center for Astrophysics], R. Gilliland [STScI], NASA, and ESA.)*

Residence:	ORION
Other name:	ALPHA ORIONIS
Class:	M1 RED SUPERGIANT
Visual magnitude:	0.55 (SOMEWHAT VARIABLE)
Distance:	430 LIGHT YEARS
Absolute visual magnitude:	–5.1
Significance:	A BRIGHT RED SUPERGIANT WITH A HUGE DUST SHELL AND THE FIRST TO HAVE ITS SURFACE DIRECTLY IMAGED.

BETELGEUSE

One of the sky's two first magnitude supergiants (the other, **ANTARES** of northern summer), Betelgeuse is so large and nearby that it presents a notable disk. Angular sizes of stars are measured by interferometry, which makes use of the way in which light radiated by different parts of the stellar surface interferes with itself. Betelgeuse was among the first to have its size determined in this way by A. A. Michelson in 1920. Modern measures give a base angular diameter of 0.045 seconds of arc, a physical diameter 650 times that of the Sun. In our Solar System, the star would extend to 2.8 AU, into the middle of the Asteroid Belt. Unlike the Sun, however, Betelgeuse has no sharp apparent "edge." Various measures, made at wavelengths in which the star's gases are particularly opaque, extend the size to 800 times solar, a radius of nearly 4 AU. The distance of 430 light years and the addition of infrared radiation from the 3600-K surface yield a luminosity 55,000 times that of the Sun.

Betelgeuse is so big that it has the honor of being the first star to have its surface imaged directly by the Hubble Space Telescope in 1996. And it looked nothing like we expected. The observations, made in ultraviolet light, showed an even bigger disk over 8 AU in radius, twice the size at which the star appears at optical wavelengths. The increase is the effect of an extended "chromosphere," a tenuous circumstellar magnetically heated atmosphere of the sort that surrounds the Sun. Moreover, offset from the center, the star showed off a large bright spot. It may be a convective bubble rising from below, but no one really knows. The spot is probably related to the star's erratic brightness variations, which take place over intervals of days, months, and years, and make it rival Rigel, supporting Johannes Bayer's seventeenth-century "Alpha" designation.

The spot may also be related to mass loss. Betelgeuse is surrounded by a complex envelope that includes a partial ring of dust with a radius about three times that of the star. It is nested inside yet another dusty ring that peaks about 50 stellar radii away. Still more dust is found from dust grains reflecting starlight up to 90 seconds of arc from the star, 12,000 AU, or 3000 times the stellar radius. And where there is dust, there must be gas, which too is found out to 1000 or so times the stellar radius.

Betelgeuse's remarkable characteristics can be explained only by high mass coupled with an advanced state of evolution. The star's central hydrogen fuel supply has run out, and as a result, its core has contracted into a hot dense state, while the exterior has swelled outward. The star ought now to be in the process of fusing its core helium into carbon and oxygen. Its luminosity and temperature can be explained only with a mass about 15 times that of the Sun, which will force the star to fuse elements all the way to iron. The iron core will collapse, and Betelgeuse will explode, most likely leaving a compact neutron star about 20 kilometers across. If it were to go today, it would be as bright as a crescent moon. If the mass is near or under the lower end of the allowed range, however, Betelgeuse may eventually become a shrunken white dwarf about the size of Earth. Even then the star intrigues. Most white dwarfs are made of carbon and oxygen, whereas Betelgeuse has enough mass to become one of the exceedingly rare neon-oxygen white dwarfs. The only way we will really know is to wait a million years.

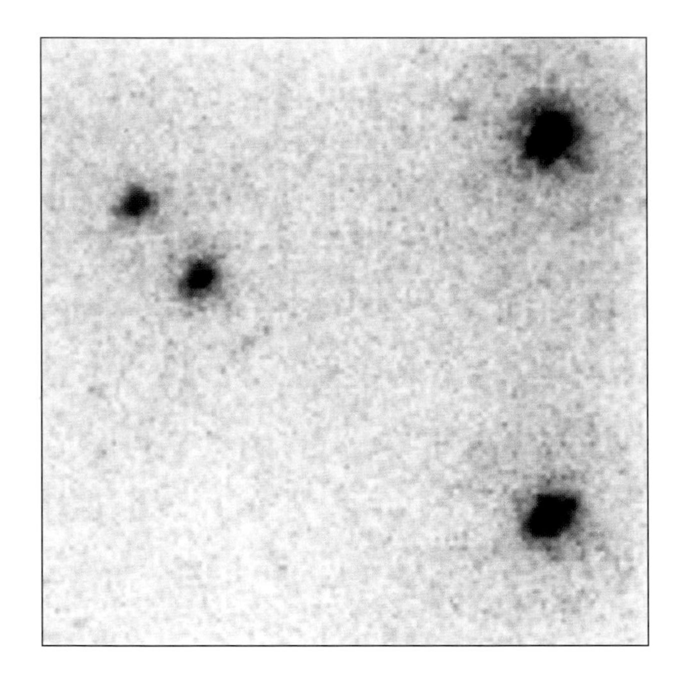

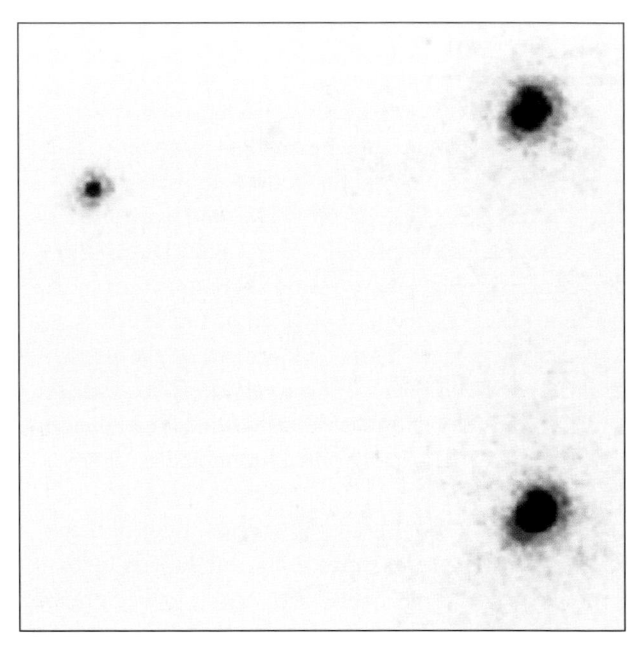

Left: The Black Widow binary, made of a star and a pulsar, appears as the lower of the two stars at left, where we see the front surface of the star violently heated by the pulsar. Right: The two have orbited 180 degrees, and the star's unheated backside is too dim to show. *(Hubble Space Telescope, A. S. Fruchter, J. Bookbinder and C. D. Bailyn, STScI, and NASA, from* ASTROPHYSICAL JOURNAL LETTERS *443 [1995]: L21.)*

Residence:	SAGITTA
Other name:	PSR B1957 +20
Class:	MILLISECOND PULSAR/NEUTRON STAR
Visual magnitude:	20
Distance:	2500 LIGHT YEARS
Absolute visual magnitude:	11
Significance:	A RAPIDLY ROTATING PULSAR THAT IS CONSUMING ITS BINARY COMPANION.

THE BLACK WIDOW

Just when you think you have found the strangest things in the Universe, something even stranger says, "Look at me." Pulsars caused a sensation when they were discovered in 1967. Imagine, we marveled, a neutron star, a star that compresses the mass of the Sun into a ball no more than 20 kilometers wide, spinning once every second or two, sending us immense blasts of radio energy as the star's magnetic axis whips by.

In 1982, astronomers were even more astonished to find a radio-beaming pulsar that was spinning at the unheard-of rate of 640 times per second! Ordinary pulsars—how blasé we become—spin fast because of the conservation of spin energy. As the iron core of an impending supernova collapses into a neutron star, it rotates faster and faster. But there is no way that such a collapse can make a pulsar spin this fast. Moreover, we observe ordinary pulsars slowing down as they age. How then could these newly found "millisecond pulsars" be spinning so very fast? Some other phenomenon must spin them back up after the pulsars' original birth and decline.

So what happens? We know that transfer of mass from one star of a double to another can alter rotation rates. Perhaps pulsars could be affected by ordinary binary companions that survived the explosion which made the pulsar. Immediately after creation, the original pulsar spins slower and slower with time, radiating its energy away, eventually disappearing from view. But now one of two things happen. If the companion is already near the pulsar, it will be drawn even closer as it loses orbital energy through its outflowing, magnetized wind. Tidally distorted, the companion passes matter into a spinning disk around the neutron star. If the companion is more distant, the pulsar waits for it to begin to evolve and expand, whereupon the same thing happens. As fast-moving matter crashes to the pulsar's surface from the disk, the old tired pulsar spins faster and faster. Reborn, it eventually spins hundreds of time per second. While the transferring action goes on, the system may appear as a "low-mass X-ray binary" rather like **MXB 1730–335**, the "Rapid Burster."

Some millisecond pulsars do indeed have binary companions. A companion reveals itself by moving the pulsar back and forth along the line of sight as the two stars mutually orbit, the motion affecting the rates at which we receive the pulses. Other pulsars, however, are clearly single. How do you spin up a pulsar that has no companion? PSR B1957 +20, which has one, solved the problem and gave us new insights into violent astrophysics. The pulsar is "visible" only in the radio spectrum. At the pulsar's location is a 20th magnitude star, which is not the pulsar, but the companion. The orbital plane is in the line of sight, and as a result, the two stars eclipse. Every orbital period of 9.2 hours, the companion passes in front of the neutron star, and for about 1 hour, the pulses disappear. From the orbit we learn that the eclipsing companion is about the size of the Sun, but has an extremely low mass of only about 2 percent solar, far below that of any respectable star.

Moreover, the companion fades in and out every orbital period in spite of the fact that the pulsar is much too small to cause an eclipse. Instead, we must be seeing a side of the star heated by the pulsar alternately facing toward and away from us. The pulsar, having been spun up by matter from the companion, is now ablating its neighbor away. The companion will soon disappear altogether, devoured by the neutron star, rendering the Black Widow pulsar single, its fate to spin for aeons without the partner that has accompanied it since birth.

17

Canopus glows low in the south above the telescope domes at the Kitt Peak National Observatory in Arizona. The star ranks second only to SIRIUS in Canis Major, which, with ADHARA, lies near the top of the picture. *(J. B. Kaler.)*

Residence:	CARINA
Other name:	ALPHA CARINAE
Class:	WHITE F0 BRIGHT GIANT OR SUPERGIANT
Visual magnitude:	–0.72
Distance:	313 LIGHT YEARS
Absolute visual magnitude:	–5.63
Significance:	THE SKY'S SECOND-BRIGHTEST STAR AND A WONDERFUL EXAMPLE OF A RARE F SUPERGIANT.

CANOPUS

How odd that the three brightest stars in the sky, **SIRIUS**, Canopus, and **ALPHA CENTAURI** are in the Southern Hemisphere, while the next three brightest (**ARCTURUS**, **VEGA**, and **CAPELLA**) are in the North. The coincidences continue, as Canopus, the second-brightest of all stars as seen from Earth, is almost exactly south (by only 36 degrees) of first-ranking Sirius. "Canopus," like "Sirius," is of Greek origin, though the name of a man.

Most northerners get their first glimpse of it while vacationing south. The astronomically aware will, after first wondering what that bright star near the horizon could be, may exclaim, "Oh! Canopus!" While Sirius is brilliant in our sky mostly because it is a mere 9 light years away, Canopus is bright because it is truly luminous. Over 300 light years distant, 36 times farther than Sirius, its 7800-K surface shines a fantastic 14,800 times brighter than our Sun. Were Canopus to be our Sun, even Pluto would be fried, and our Earth would have to be thrice Pluto's distance from the Sun just to receive the proper heating. Though Canopus's spectrum has long suggested giant status, the enormous luminosity finally calculated from a good parallax puts it clearly into the realm of the supergiants. The star is large enough to have had its angular diameter measured at just over 0.006 seconds of arc, giving it a true diameter of 0.6 AU (the same value derived from temperature and luminosity). Some 65 times larger than our Sun, this wonderful star would extend 75 percent of the way to the orbit of Mercury.

Canopus is one of the only stars that has been observed across nearly the whole electromagnetic spectrum. As an F star not too much hotter than the Sun, most of the light from its surface is radiated in the visible and in the soft ultraviolet that penetrates the Earth's atmosphere. However, it is also a strong source of energetic X-rays that cannot possibly be produced by a cool body at only 7800 K. Canopus must also have a magnetically heated corona somewhat like the one surrounding the Sun. To produce the observed X-rays, however, Canopus's corona must be very hot, somewhere around 15 million K—ten times hotter than the solar corona. A stellar corona is heated by magnetism generated by a combination of the star's rotation and the seething up and down convection in its outer layers. To have such strong coronal radiation, Canopus's magnetic dynamo must be efficient indeed. The star's rotation is not observed, however, possibly because its rotation pole points toward us. Remarkably, Canopus is also observed with radio telescopes, this radiation produced by electrons accelerated in the powerful coronal magnetic fields.

Canopus is an evolved star, one that has ceased hydrogen fusion in its core and should now be fusing helium into carbon. These internal changes have expanded the star to its current proportions. Most such stars are red giants or supergiants. Canopus, however, is one of a rare breed of yellow supergiants. We do not know whether it had once been a red giant or supergiant (like **BETELGEUSE**) and is currently heating and becoming bluer as internal adjustments continue, or if it is cooling and becoming redder. But it probably began life as an 8- or 9-solar-mass star, just under the limit at which it would develop an iron core and explode. Canopus, part of an extended family of stars in the "Scorpius-Centaurus association" that were more or less born together, will someday probably become one of the Galaxy's most massive white dwarfs, perhaps one massive enough to fuse its carbon to become a rare neon-oxygen white dwarf—a possible ending for Betelgeuse as well.

Near the bottom of the picture, bright Capella and her triangle of three "Kids" (EPSILON AURIGAE the one nearest Capella) shine through the structure of the 2.3-meter telescope of Steward Observatory on Kitt Peak in Arizona. *(J. B. Kaler.)*

Residence:	AURIGA
Other name:	ALPHA AURIGAE
Class:	YELLOW GIANT G0/G8 DOUBLE STAR
Visual magnitude:	0.08
Distance:	42 LIGHT YEARS
Absolute visual magnitude:	–0.47
Significance:	THE FIRST MAGNITUDE STAR CLOSEST TO THE NORTH CELESTIAL POLE AND A RARE YELLOW-GIANT DOUBLE.

CAPELLA

Helped by atmospheric absorption, Capella rises above the northeastern horizon emitting a warm golden light. Sixth-brightest star in the sky, just barely behind ARCTURUS and VEGA, Capella ranks third in the Northern Hemisphere. The northern counterpart to ACRUX, it is the closest first magnitude star to the north celestial pole and circumpolar to anyone living above 44 degrees north latitude. The luminary of Auriga, the Charioteer, "Capella" is from old Latin, and means "the She-Goat"; Auriga is commonly depicted carrying her under his arm. Nearby is a pretty asterism of three stars, "The Kids," one of which, EPSILON AURIGAE, is one of the strangest variables in the sky. At a distance of only 42 light years, Capella has a deceptive luminosity of 133 Suns because the star is actually a close double. The two components, a few hundredths of a second of arc apart, are just close enough to be inseparable to the eye through an Earth-based telescope. The naming of the stars is a mess. The components of a double are usually called "A" and "B," and Capella's frequently are. However, before Capella proper was known to be double, "B" was used for another nearby star. When Capella "A" turned out to be double as well, its components were named Capella Aa and Ab. Whoever said astronomy is not logical? Confusing, maybe, but logical. "Capella B" is not part of the system and simply lies in the line of sight, as do "C" through "G," leaving only little tenth-magnitude "H." Also a double made of two red dwarfs, Capella H orbits Capella A a good fraction of a light year away over a period of thousands of years. The whole system is thus quadruple.

Studies with spectrographs and interferometers (the devices used to measure stellar diameters) determined the orbital parameters long ago. Capella Aa and Ab orbit every 104 days. The brighter (Aa) is a G8 giant a bit cooler than the Sun, while the fainter (Ab) is a somewhat warmer G0 giant. The cooler giant, 82 times more luminous than the Sun, has a mass of 2.69 solar and a radius 14 times the Sun's. The dimmer has a luminosity 51 times the Sun's, and a mass and radius of 2.56 and 11 solar, respectively. These two magnificent giants are separated by about the distance between Venus and the Sun. A resident on a "Jupiter" ten times farther out would see two "Suns" about half a degree across (similar to the Sun in our own sky), separated at maximum by some 6 degrees, one setting right behind the other.

Both Aa and Ab are dying, evolving from nearly identical stars that were once class A dwarfs much like Vega is today (doubles like this abound, EPSILON LYRAE consisting of two of them). The higher-mass star always evolves first. Aa, the more massive giant, is believed to have been a red giant and to have stabilized by fusing helium to carbon in its core. Less massive Ab, however, has not reached this state and may still have a shrinking helium core that has not yet fired up. Because it is more developed, Aa is also rotating more slowly. Astronomers in the far distant future will see a pair of nearly identical orbiting white dwarfs.

The pair, a source of X-rays, have some characteristics of the class of double stars epitomized by RS CANUM VENATICORUM, whose interactions produce activity similar to that found in the Sun's chromosphere, and even starspots. Astronomers cannot nail down which star is responsible for the X-rays, however, Aa or Ab (or both). More rapidly rotating Ab should be the more active, yet single stars like DENEB KAITOS that are similar to Aa are also X-ray sources. The "She-Goat" looks blithely downward, challenging us to learn her secrets.

The spiral galaxy NGC 4013 stands in for our own Galaxy to show the relative positions of the Sun and CD–38°245. *(C. Howk [JHU], B. Savage [University of Wisconsin], N.A. Sharp [NOAO], WIYN/NOAO/NSF.)*

Residence:	SCULPTOR
Other name:	NONE
Class:	CLASS G OR K GIANT
Visual magnitude:	12.0
Distance:	5000 LIGHT YEARS
Absolute visual magnitude:	ABOUT +1
Significance:	THE MOST METAL-POOR STAR KNOWN.

CD–38°245

From 1852 to 1859, Friedrich Argelander and his assistants at the Bonn Observatory in Germany mapped the northern sky to about ninth magnitude, compiling the position and brightness of 324,000 stars. The catalogue, the *Bonner Durchmusterung* ("Bonn Survey"), numbered the stars sequentially from the vernal equinox (where the Sun crosses the celestial equator in March) in degree-wide strips of declination (a coordinate akin to latitude). Stars are still commonly known by their "BD" numbers. Between 1885 and 1930, the work was continued in the Southern Hemisphere at the Cordoba Observatory in Argentina, the southern catalogue known as the *Cordoba Durchmusterung*, or "CD." Our star, CD–38°245, is the 245th star within the band that lies between 38° and 39° south "declination." At 12th magnitude, CD–38°245 lies right at the limit of the Cordoba Survey in the obscure modern constellation Sculptor near the South Galactic Pole, which is defined by the perpendicular to the Milky Way.

Classification—whether OBAFGK or M—from stellar spectra involves ordinary stars like the Sun and its kin, which inhabit the Galaxy's disk and have similar high-metal compositions. Unlike the vast majority of stars, CD–38°245 has a very low abundance of metal atoms, which dramatically changes its spectrum and makes it difficult to classify. Its yellow-orange color and surface temperature of 4700 K would make it class G, somewhat to the cool side of the Sun. But metal-deficient stars have temperatures too hot for their apparent classes, so the best comparison would be to cooler class K. Based on its spectrum, the star was believed to be an evolved giant. Were it a main sequence dwarf star like the Sun, it would be close to us, its distance easily measured by the Hipparcos parallax satellite. That the parallax is unmeasurable clearly confirms its evolved giant status. From an assumed absolute magnitude of +1 (40 times brighter than the Sun), the star lies 5000 light years away.

The difficulty of classifying this star is what makes it so important. From its spectrum, the iron content is measured to be but a ten-thousandth that of the Sun! Perhaps not suprisingly, such records are steeped in controversy. Several other stars have similarly strange compositions, and one other has been uncertainly touted as having ten times fewer metal atoms. Nevertheless, our CD–38°245, if nothing else, stands in for the set of extremely metal-poor stars.

Why are these important? CD–38°245 is almost "straight up" relative to the plane, or disk, of our Galaxy, and is 5000 light years into the sparsely populated Galactic halo where the oldest stars reside. The standard (much simplified) view of the evolution of the Galaxy is that through their winds and explosions, stars sent heavy elements into the star-forming gas of interstellar space. Therefore, recently-born stars should have more metals from earlier generations of stars, and old stars should have fewer. That the lowest-metal stars are found in our Galaxy's halo helps show that the halo is the oldest part of the Galaxy. CD–38°245, with the lowest metal content, must be among the very oldest of stars.

The Big Bang, the creation event that formed the Universe, is thought to have delivered only hydrogen, helium, and a bit of lithium to the cosmos. As a result, the first stars should have had no significant metals at all. We cannot find them. CD–38°245, therefore, not only holds a record, but points to deep mystery. Where is the first stellar generation?

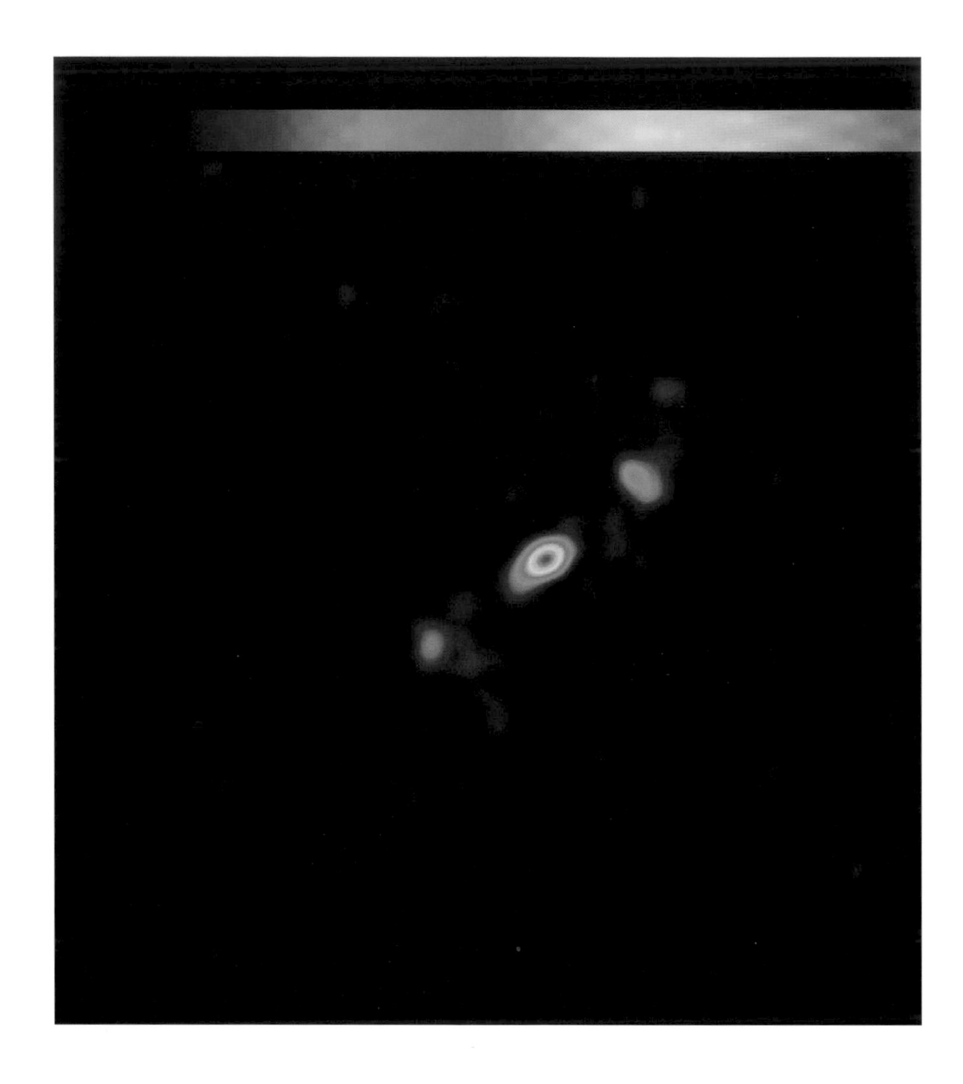

The great 1976–86 outburst of CH Cygni, during which it became visible to the naked eye, was associated with oppositely-directed jets visible in radio radiation. *(Very Large Array, National Radio Astronomy Observatory, courtesy of R. Taylor and M. Crocker.)*

Residence:	CYGNUS
Other name:	HD 182917
Class:	RED M7 GIANT SYMBIOTIC STAR
Visual magnitude:	6 TO 8
Distance:	875 LIGHT YEARS
Absolute visual magnitude:	ABOUT –1 TO +1
Significance:	THE BRIGHTEST SYMBIOTIC STAR.

CH CYGNI

Astronomers name variable stars with single or double Roman letters, a holdover from an old nomenclature system developed by Johannes Bayer. From its name, CH Cygni is clearly variable. And how oddly so. The star's variation is linked to duplicity and mass transfer, an ALGOL gone mad. The principal star is a red giant. Red M giants make a grand display of titanium oxide (and other) molecules that can form only under low-temperature conditions. However, a small class of red giants adds spectral emissions of hydrogen and of wildly excited ions typical of planetary nebulae (such as NGC 7027) that have hot central stars. At 3000 to 4000 K, however, the red giants are far too cool to produce such emissions. Moreover, these stars often appear to have far too much blue light. The very different high- and low-temperature features simply could not be reconciled with each other. Without much understanding of these stars—or of biology for that matter—astronomers named them "symbiotic stars," the idea being one distinct property living in some kind of coexistence with another.

Red giants, including those with symbiotic features, commonly display brightness variations that range from erratic meanderings to full-blown MIRA pulsations. However, the symbiotics take the variations vastly farther. The emissions and the "blue continuum" might vary quite independently of those of the red giant. More dramatically, the high-temperature features can suddenly brighten into some kind of "outburst" state in which the stars can remain for a considerable period of time. Theorists fought over it. One camp argued that the stars were single and perhaps producing a planetary nebula. Others suggested stellar duplicity might be at work.

CH Cygni epitomizes this stellar wonder. Outwardly, it is a cool M7 giant with a temperature around 3500 K. Not quite 900 light years away, the red giant erratically hovers around seventh or eighth magnitude, with a typical luminosity 50 times solar and a measured diameter about 300 solar, big enough to occupy the orbit of Mars. Typical of symbiotics, it radiates emissions of hydrogen, helium, oxygen, and other elements. Every few years, it erupts for long periods and has reached fifth magnitude, making it possible to see with the naked eye. Even in recent times astronomers have fought fiercely over it. Again, some thought it was single, while others thought they observed eclipses with a 15-year period, indicating duplicity. Others suggested yet a triple system with two eclipse cycles with 2- and 15-year periods.

The most likely picture of CH Cygni—as for all symbiotics—is a combination of a giant and a smaller star, usually a compact white dwarf. CH Cygni's huge, extended red giant produces a wind that is drawn to a white-dwarf companion. The wind settles into a disk, an accretion disk, from which the entrapped gas heats and falls to the white dwarf's surface. The high temperatures produced excite the wind and make it radiate the high-excitation emissions, even X-rays. The disk is very unstable and may suddenly dump itself onto the white dwarf, the star suddenly erupting to naked-eye visibility, variations in the disk probably causing false "eclipses." As the disk empties, massive jets shoot out from the poles, and the star is then more or less quiet for awhile. The red giant will probably produce a planetary nebula, one that will be powerfully influenced by the tiny companion. And what sport it is to watch it all from the backyard.

Chi Cygni (see also #86), a class S MIRA-type variable star, is rich in zirconium oxide. S stars and their carbon-star progeny create most of the zirconium in the Universe, without which beautiful zircons (zirconium silicate) like these would be impossible. (© *theimage.com.)*

Residence:	CYGNUS
Other name:	HR 7564
Class:	S6 GIANT
Visual magnitude:	4(MAXIMUM BRIGHTNESS)
Distance:	300 LIGHT YEARS
Absolute visual magnitude:	ABOUT –1 MAXIMUM
Significance:	MIRA VARIABLE AND THE BRIGHTEST S STAR.

CHI CYGNI

While some stars that make major parts of their constellations simply vary (**ALGOL** a case in point), two have the effrontery to disappear! The more famed is **MIRA**, whose comings and goings make Cetus change its form. The other, Chi Cygni, a more obscure Mira-type variable, lies in the staff of the Northern Cross.

"Miras" are huge stars the size of the orbits of Earth and Mars. In the astronomy trade, they are a subset of the awfully-named "asymptotic giant branch," or "AGB," stars. These stars, more rationally called "second-ascent giant stars," are in their last throes and have contracting carbon-oxygen cores surrounded by fusing shells of helium and hydrogen. Radiating thousands of solar luminosities, they become larger, cooler, and redder, until they pop planetary nebulae and expose their white-dwarf cores. As AGB stars grow, they become unstable and begin to pulsate, and, near their ends, become the famed "Miras."

Chi Cygni is among the more extreme of the breed, its magnitude varying from an easily-seen 3.5 to a miserable 14 and back over a period of 400 days. At its brightest, it is visually 15,000 times fainter than at its dimmest! (Such huge variation is deceptive, however, and is caused more by the chilling of the star to lower temperature, where it radiates invisible infrared light.) At its maximum luminosity, it radiates about 3000 times the energy of the Sun and is somewhere around 300 times the solar diameter, making it twice the size of the orbit of Mars.

But curiously, these impressive statistics are not what make Chi Cygni special. Even great "Miras" like Chi are rather a dime a dozen. Chi Cygni also happens to be the sky's brightest S star. As far back as the time of the American Civil War, Father Angelo Secchi had separated the cool red stars into those with titanium oxide absorptions and those with carbon. In 1890, William Pickering classified the stars A through Q mostly on the basis of their hydrogen absorptions. At that time, Father Secchi's two classes became M and N, **R LEPORIS** among the brightest of them.

In the 1920s, a curious hybrid was recognized that displayed absorptions of zirconium oxide. These stars were assigned to a new class, "S." As few others, S stars reveal alchemy in action. The AGB stars within a small mass range contain overlapping convection cells that can take the carbon made by the fusion of helium and loft it to the surface, creating the deep red carbon stars of class N. Ordinarily, a star has twice as much oxygen as carbon. When carbon dominates, it grabs all the oxygen, leaving none for titanium. But when the growing carbon about equals oxygen, the few remaining oxygen atoms catch those of zirconium, which it likes much better than titanium, and for a brief shining moment zirconium oxide blooms.

But there is much more. Along with carbon, zirconium is also increasing in abundance. Theoreticians know the element to be created by the capture of neutrons onto atoms of yttrium, the zirconium at the same time transforming to niobium. The star is making heavier atoms in its helium-burning shell and bringing them to the surface. Proof of such action is that S stars (as well as those of other cool classes) contain absorptions of rare technetium, an element so radioactive that none exists on Earth. Almost all of our zirconium—and many similar elements that are used heavily in industry—once came from earlier generations of such stars. When we view Chi Cygni, we know it is making yet more heavy elements. Lofted into the cosmos by winds, perhaps they will be used by some civilization of the far-distant future.

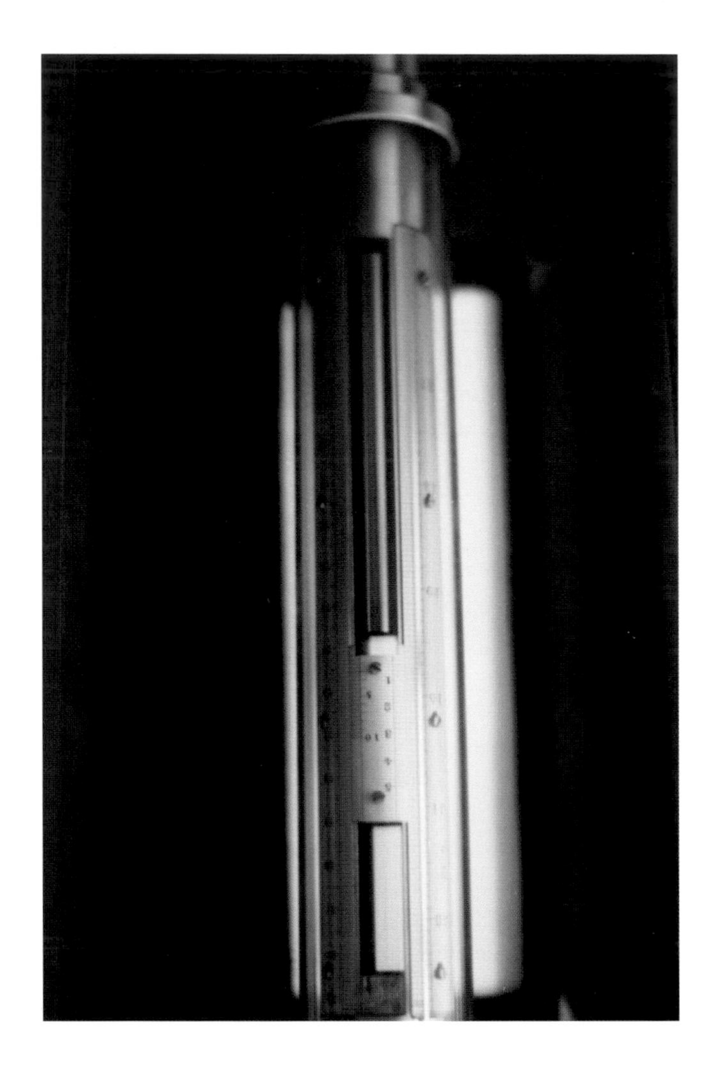

Mercury, a rare and dangerous silvery liquid metal used in older thermometers and barometers (like the one pictured here), can approach the abundance of iron in the outer layers of stars like Chi Lupi. *(J. B. Kaler.)*

Residence:	LUPUS
Other name:	5 LUPI
Class:	B9 SUBGIANT PLUS A2 DWARF
Visual magnitude:	3.95
Distance:	205 LIGHT YEARS
Absolute visual magnitude:	–0.05
Significance:	THE "PATHFINDER" MERCURY-MANGANESE STAR.

CHI LUPI

Squeezed between Scorpius and the Centaur, Lupus (the Wolf) lies in wait at the edge of the Milky Way. Among its treasures is a modest fourth magnitude star just below the Scorpion's head, Chi Lupi. The spectrograph reveals that Chi is double. Separated by half the distance between Mercury and the Sun, the pair whirl around each other every 15 days. The brighter, slightly evolved B9 star of about three solar masses emits 80 percent of the light, not enough to overwhelm the smaller two-solar-mass A2 main sequence companion. Chi's uniqueness lies in its amazing atomic abundances (As a result, it was adopted as a Hubble "Pathfinder" star.). Dominated by hydrogen and helium, the Sun is the standard reference for stellar chemical composition. Heavy elements, so common on Earth, are rare. For example, the Sun contains 1 atom of iron for every 32,000 hydrogen atoms and but 1 of gold for every 100 billion hydrogen atoms.

The exceptions to the solar rule help us learn about how other stars work. Odd abundances can be the result of nuclear reactions and mixing from below, which produce carbon stars like **R LEPORIS** and **IRC +10 216**. Other stars, exemplified by **DENEB KAITOS**, were contaminated by evolving carbon-rich companions. The components of Chi Lupi, however, are very different. They are overabundant in heavy stuff while depleted in other elements. The dimmer of the pair is a "metallic-line star"; such stars are typically enriched by factors of 10 or so in copper, zinc, strontium, zirconium, barium, and europium, while at the same time depleted in calcium and scandium. The brighter B component is an even odder "mercury-manganese star," in which platinum and gold are up by tens of thousands, and mercury has over 1 million times the solar abundance! Yet aluminum and calcium are notably underabundant.

Chemical-building by nuclear reactions cannot explain such odd chemical adundances. The metallic-line stars are all warm classes A and F ("Am-Fm stars" in the jargon), while the mercury-manganese stars are hotter class B. All rotate slowly, and a good fraction are double, gravitational coupling between the pairs slowing the stars' rotations. In classes F through B, where temperatures go from 6000 to 20,000 K, circulation by convection also settles down. As a result, the stellar atmospheres are very quiet. In these odd stars, epitomized by Chi Lupi, some atoms fall under the action of gravity. Other atoms that absorb light in the range of colors where the star is very luminous are pushed outward by the pressure of radiation. Consequently, different elements physically separate. A small wind blowing away the outer surface layers (in addition to the circulation of gases caused by some rotation) then yield abundance patterns that depend on temperature. Stir in a magnetic field, as does **COR CAROLI**, and things really get crazy. A large percentage of stars fall into these chemically peculiar classes; there may be no "normal" A stars at all. Chemical composition is one of the most critical stellar properties we can measure and sets the stage for knowing how our Galaxy evolved. Chi Lupi teaches us that we must be careful about interpreting stellar chemical abundances until we thoroughly understand all the physical processes at work.

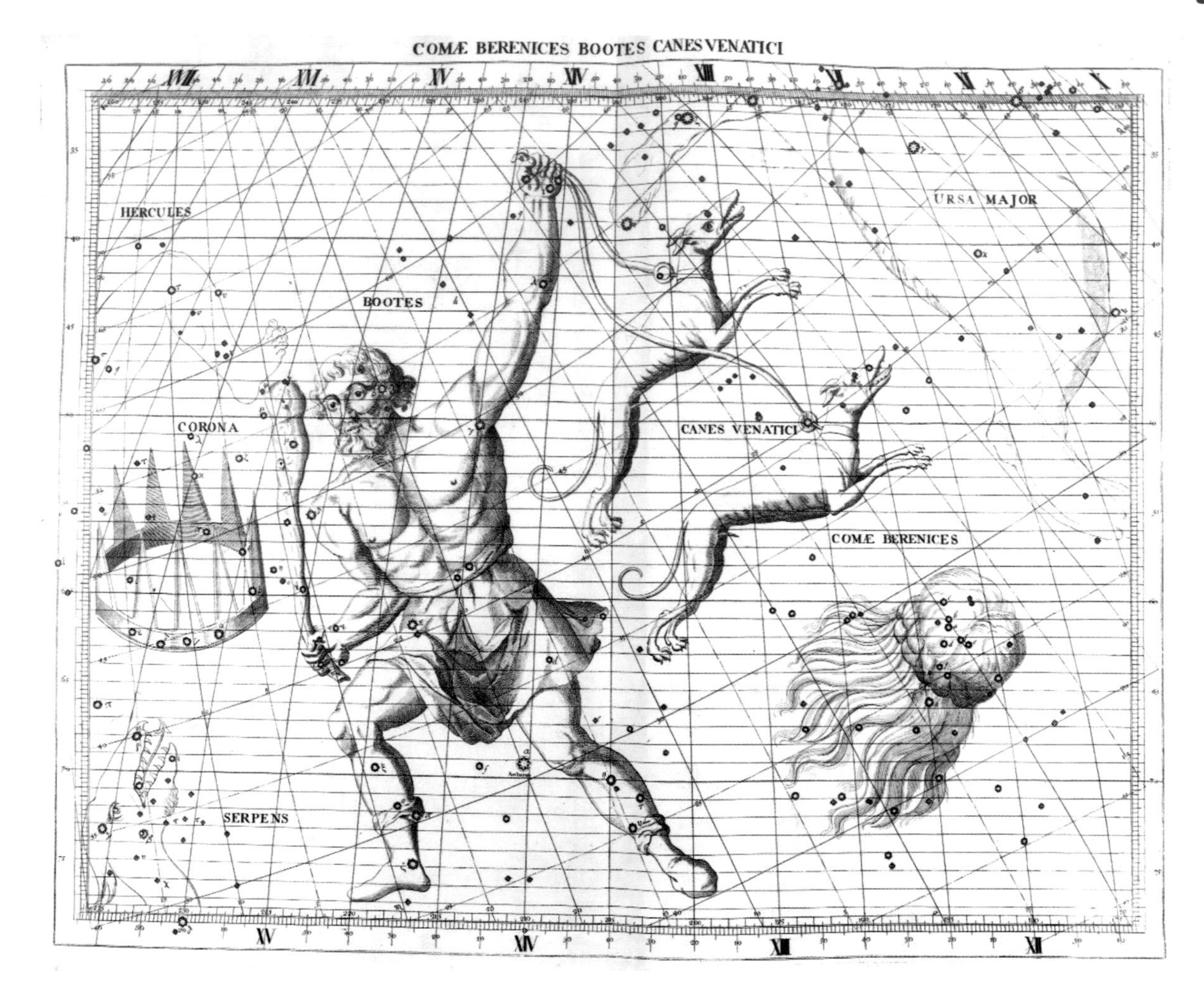

Cor Caroli appears as the bright star in the collar of the "southern dog" of the modern constellation Canes Venatici, the Hunting Dogs. *(John Flamsteed,. 1729.* Atlas Coelestis. *Courtesy Rare Book Room and Special Collections Library, University of Illinois.)*

Residence:	CANES VENATICI
Other name:	ALPHA-2 CANUM VENATICORUM
Class:	A0 MAGNETIC STAR
Visual magnitude:	2.90
Distance:	110 LIGHT YEARS
Absolute visual magnitude:	0.26
Significance:	ODD CHEMICAL ABUNDANCES IN A STRONG MAGNETIC FIELD.

COR CAROLI

Cor Caroli is the modest luminary of the simple constellation Canes Venatici, the Hunting Dogs, created in the 1600s by the astronomer Hevelius from stars south of the Big Dipper's handle. It is a double with third and fifth magnitude components separated by 20 seconds of arc, making it a fine sight in a small telescope. Following tradition, the western star is called "Alpha-1 Canum Venaticorum," while the eastern—the brighter and the focus of our attention—is "Alpha-2." Though Alpha-1 is a true binary companion, it is just an ordinary class F dwarf with no bearing on the issue of Alpha-2, which stands out as a magnetic variation on the theme of chemically peculiar stars such as **CHI LUPI**.

To understand Alpha-2, we need our Sun. Acting together, solar rotation and convection create the solar magnetic field. The two then work against each other, the rotation wrapping the field into dense powerful ropes, which break through the surface to inhibit convection and form dark sunspots. Magnetic fields are found and analyzed in the Sun and other stars by the "Zeeman effect" (named after the Dutch physicist Pieter Zeeman). As light flows through a relatively dense gas, its atoms absorb at particular colors, producing the narrow gaps seen in stellar spectra. If the absorbing gas is in a magnetic field, the lines will be split apart with the separation of the components proportional to the field strength. In the 1940s, astronomers began to find the Zeeman effect not just in sunspot spectra, but in the spectra of class A stars that also have peculiar chemical compositions. One of their number, Cor Caroli, stands in for the whole set of magnetically peculiar stars. They range through both classes A and B (7400 to 23,000 K) and are called "Ap-Bp" stars, with Cor Caroli, a two-solar-mass class A star right in the middle.

These warm magnetic stars are in a league with the metallic-line and mercury-manganese stars epitomized by Chi Lupi. They too are slow rotators and have huge over-abundances of some elements, notably iron (and those of similar weight) and "rare-earth elements" like europium. Separation of the elements—the sinking of some, the raising of others—by the competing forces of gravity and the pressure of radiation is surely the cause. But here, we also see magnetic fields that range in strength from a few hundred times that of the Earth through about 5000 times for Cor Caroli and, at the extreme, tens of thousands of times terrestrial.

As we watch, the spectra of the altered elements change dramatically with time, showing that various elements are concentrated into particular zones on these stars, most likely starspots that are vaguely similar to those found on the Sun, but vastly longer-lived. As the spots move in and out of view, the apparent chemical composition of the star changes. The magnetic fields are so strong that they change the up- and down-drift patterns of different kinds of atoms. Why some stars are magnetic and others not, however, we do not know. Stars like Cor Caroli may be the progenitors of magnetic white dwarfs such as **EG 129**. To understand the entire flow of stellar evolution, including its peculiarities, and even to learn something of the origin of the Sun's own magnetic personality, we need to learn the reason.

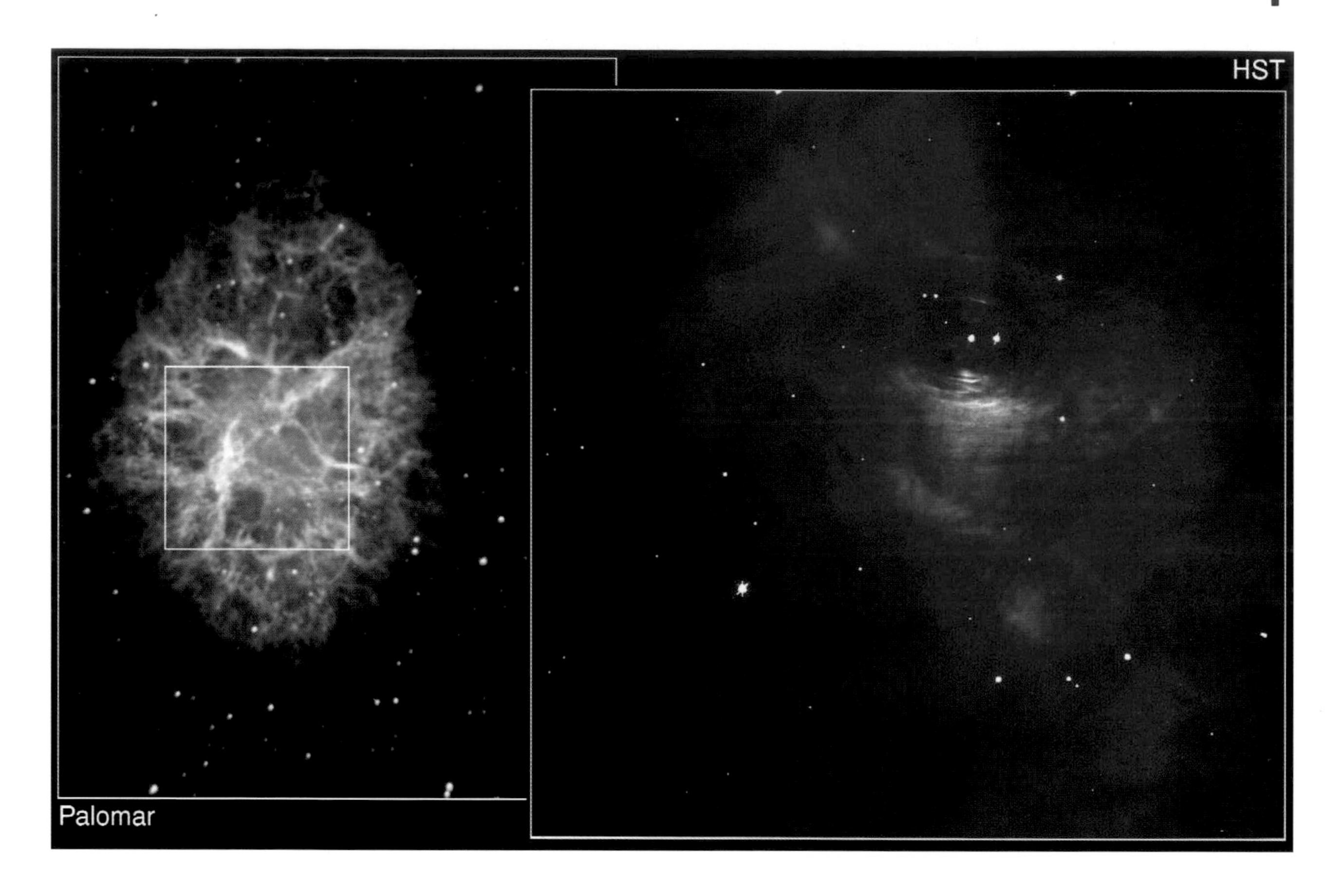

The powerful wind of the Crab Pulsar (right), the neutron-star remnant of the Crab Nebula (seen at left in a Palomar image), clearly disturbs the surrounding gas. *(Hubble Space Telescope, J. Hester and P. Scowen [Arizona State University], STScI, and NASA.)*

Residence:	TAURUS
Other name:	NP 0532; SUPERNOVA OF 1054
Class:	SUPERNOVA AND PULSAR
Visual magnitude:	OF PULSAR: 16
Distance:	6500 LIGHT YEARS
Absolute visual magnitude:	5
Significance:	FAMED SUPERNOVA REMNANT AND THE FIRST OPTICAL PULSAR.

CRAB NEBULA

July 5, 1054, dawned peacefully; except that Taurus displayed a brilliant star where none had been seen before. Recorded by both Chinese scholars and native North Americans, it shone brilliantly enough to rival Venus at her best. Fading from naked-eye view after a year, it was not heard from again for almost seven centuries, until the eighteenth-century English astronomer John Bevis found a fuzzy nebulosity near Taurus's southern horn. Almost 900 years after the event, astronomers made the connection that the glowing cloud was the rapidly expanding ejecta of an exploded star. Of the stars seen inside the Crab Nebula, one has no spectrum lines, suggesting some 70 years ago to Walter Baade and Fritz Zwicky of Mount Wilson Observatory that it is the dense collapsed remains of the detonation. But how to prove it?

Switch now to central England in 1967. Jocelyn Bell Burnell has entered a unique radio observatory—one designed to observe fast changes in celestial objects—to check on the night's work. There, she finds a record of odd pulses that come and go over the next days and weeks, but when they do appear, they are right on schedule according to an exact period of 1.33701. . . seconds. She had found an amazing natural clock residing in the modern constellation Vulpecula.

Many more of these pulsars have since been discovered. One of them, with a period of only 0.033 seconds, resides inside the Crab Nebula. Optical monitoring with fast systems showed Baade and Zwicky's old star to be the one, flashing 30 times per second—faster than the eye can see. The only reasonable explanation of such exactness is rotation. But to rotate so fast, a star must be small—only a few tens of kilometers across—and very dense. Pulsars must be neutron stars in which the chemical elements are crushed to neutrons at a density 100 trillion times that of water. An average cubic centimeter of the little star would weigh 100 million tons!

Observation alone tells us little; we need theory. Pulsars are the end-products of massive stars like **ANTARES** and **MU CEPHEI**. Over millions of years, successive stages of nuclear fusion in such stars grow iron cores that violently collapse and compress to neutrons. When maximum density is reached, a newly formed 20-kilometer-wide neutron star bounces. The resulting shock wave blows the rest of the star apart, and a supernova blooms into the night sky. The collapse causes the neutron ball to spin with incredible speed and condenses a magnetic field to immense proportions, the Crab's field 1 trillion times that of Earth. Like our planet's field, the magnetic axis is tilted against the spin axis. The intense magnetism causes radiation to beam along the field axis, which is tilted and flops around as a result of the spin. If Earth is in the way, we get a pulse and see a pulsar.

The Crab Pulsar is young and among the most energetic known, having sufficient spin energy to generate even gamma-ray pulses. Over time, the radiation will sap the rotation, and the Crab Pulsar will spin ever slower, the phenomenon actually seen by comparing the little star against accurate clocks. Eventually, it will stop radiating at high energies and will be visible only in the low-energy radio. By then, the Crab Nebula will have long-since added another load of heavy stuff to interstellar space, leaving its pulsar roaming the Galaxy all alone like many of its far-away kin.

25

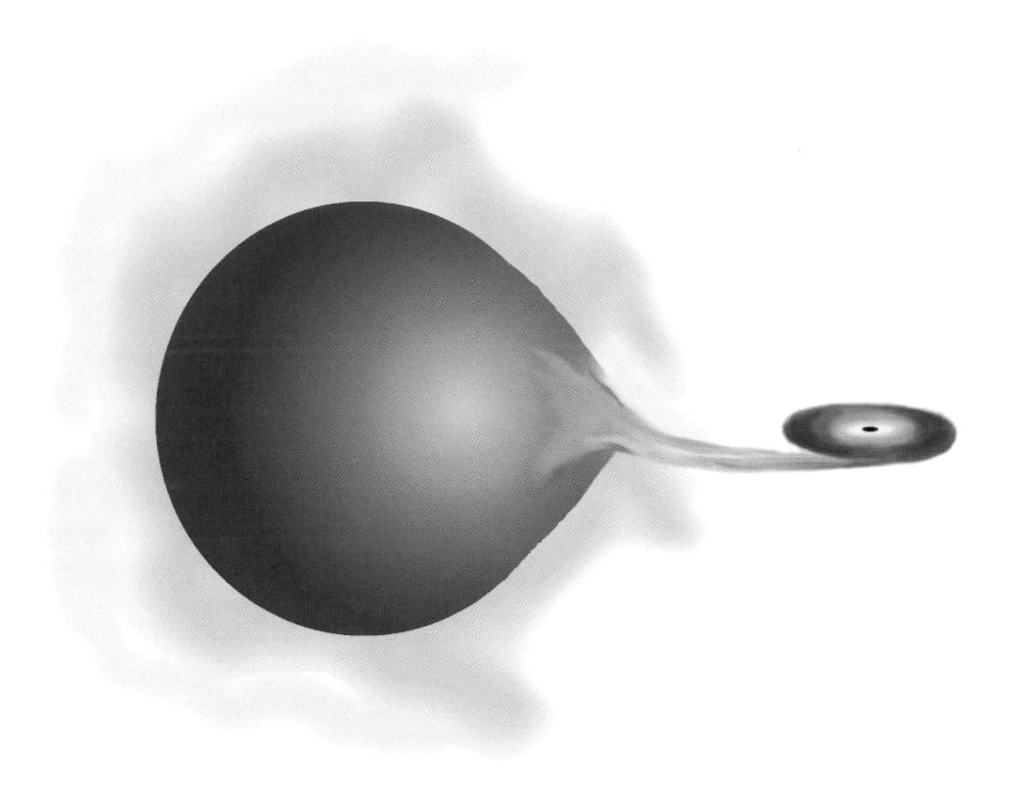

The inflowing gas surrounding the black hole at the heart of Cygnus X-1 is so hot it radiates X-rays as it prepares to disappear forever. *(NASA/CXC.)*

Residence:	CYGNUS
Other name:	V1357 CYG
Class:	O9 SUPERGIANT AND BLACK HOLE
Visual magnitude:	8.8
Distance:	8000 LIGHT YEARS
Absolute visual magnitude:	–9.3
Significance:	A WONDERFUL BLACK-HOLE CANDIDATE.

CYGNUS X-1

The night launch of Apollo 17 aboard the mighty Saturn V was said to resemble sunrise. Yet the return lunar module that squirted the astronauts home from the Moon was tiny. Escaping the Earth's gravity requires a speed of 11 kilometers per second, to leave the Moon a mere 2. This "escape velocity" depends on the escaped body's size and mass, smaller radius and larger mass both increasing the critical limit, the low lunar mass more than offsetting its quarter-Earth diameter. Escaping the surface of the Sun (in thought only, thanks) would take a speed of 600 kilometers per second, the result of its vastly larger mass.

While the Earth's escape velocity is constant, the Sun's is not. When the solar nuclear fire ceases, the Sun will expand and lose its outer layers, the remainder—what is now the nuclear burning core and its environs—becoming a shrunken white dwarf. Though losing half its mass, the Earth-sized dregs will have an average density of 1 ton per cubic centimeter, and the surface escape velocity will climb to 5000 kilometers per second. And yet because atomic particles—protons, neutrons, and electrons—take up so little space, there is still immense squeezing room. White dwarfs are very stable. But triple a typical white dwarf's mass, and it would collapse to the next stable state, in which the particles are jammed together to make neutrons. Such "neutron stars" abound; the most famed, the **CRAB NEBULA PULSAR**, was produced by the supernova of 1054. Don't try to leave. Taking off from a star with a diameter of 20 kilometers would require your rocket to accelerate to 200,000 kilometers per second, two-thirds the speed of light!

Were the star compressed to a diameter of 10 kilometers, the escape velocity would hit light-speed. Now not only can't you leave, but neither can anything else, including light. The "star," now so dense it cannot support itself at all, collapses as a fabled "black hole" and, except for its gravity, disappears. A more accurate explanation, which comes from Einstein's theory of relativity, views the hole as a puncture in spacetime.

Do such bizarre things exist? Imagine two massive stars (greater than ten solar masses each) born in mutual orbit, not unlike **GAMMA-2 VELORUM**. The more massive evolves first and runs its core to iron, which collapses and explodes as a supernova. The old core might remain as a neutron star. However, if the original star is in the right mass range (probably above 40 times solar), the core might collapse past the neutron star state and fall into the hole. As the companion evolves and swells, tides raised by the condensed body may stretch it out into an oval, from which mass flows into a disk around the little remnant that heats so much it radiates X-rays. Nothing else we know of can produce such quantities of energetic radiation.

There are stars that fit such a theoretical scenario. Cygnus X-1 appears as a class O supergiant. Unlike normal O stars, it radiates powerful X-rays. Its continually shifting spectrum tells us it is revolving around an invisible companion. Analysis of the supergiant's motion shows it to carry a mass 30 times solar, as expected from 400,000 solar luminosities! The companion's mass is estimated at 16 solar, far above the theoretical three-solar-mass limit for a neutron star. The "something" can only be a black hole. While there are over a dozen other black-hole candidates, most having more common kinds of stellar companions, Cygnus X-1 beautifully tells us that black holes live. You can see this one—at least the supergiant—from your backyard with a small telescope. Just don't try to get out of one.

Delta Cephei is the prototype of Cepheid variable stars that are used to determine the distances to other galaxies. The variation of a Cepheid found in the great spiral galaxy M100 (56 million light years distant) is shown across the top. *(Hubble Space Telescope, W. L. Freedman, Observatories of the Carnegie Institution of Washington, STScI, and NASA.)*

Residence:	CEPHEUS
Other name:	27 CEPHEI
Class:	F TO G SUPERGIANT
Visual magnitude:	4.3 TO 3.5
Distance:	980 LIGHT YEARS
Absolute visual magnitude:	–3.1 TO –3.9
Significance:	THE PREMIER CEPHEID VARIABLE.

DELTA CEPHEI

Speed up time by a million. Set amidst the steady stars would be hosts of blinking fireflies. Some flicker madly several times per second, while others appear every few seconds and then descend into darkness. Delta Cephei, one of the flickerers, is among the most famed and most important of all stars. In 1784, the English amateur astronomer John Goodricke, who had found the regularity of ALGOL, watched Delta Cephei change by one magnitude over a period of five days. Much later, observations showed the star to be so remarkably regular that it was thought for a time to be an eclipsing double. But no orbit would fit, and it became obvious that this single star was cyclically varying between apparent magnitude 3.5 and 4.3 over a period of exactly 5 days, 8 hours, 47 minutes, and 32 seconds.

Just beating Goodricke's discovery was one made by his friend Edward Piggott of similar variations by Eta Aquilae. Nevertheless, Delta Cephei gained ascendancy, and a whole class of thousands of such variables became known as "Cepheids." Though generally behaving alike, the Cepheids' periods range from 1 day to over 50 days. Like Delta Cephei proper, all rise quickly to maximum and then slowly fall back to minimum. And though there are many imitators, the "classical" Cepheids like Delta are all dying, but brilliant, class F or G supergiants. Delta, at a great distance of 1000 light years, is the closest of them, its luminosity changing between 1500 and 3000 times that of the Sun.

Delta Cephei and its ilk vary by pulsation. Like a child on a swing, this great star, some 40 times the solar diameter (so large it can be detected as a disk), cannot find a place for itself. In a fierce tug-of-war between outward-pushing gas pressure and inward-pulling gravity, it alternately expands and contracts, changing its spectral class and surface temperature as it goes. Greatest brightness depends both on temperature and radius, and it takes place not when the star is largest or smallest, but when (from Doppler measures) it is expanding the fastest; similarly, it is faintest when so contracting.

Like our swinging tyke, the oscillations would cease were it not for a "pusher," a layer well below the surface in which helium is becoming doubly ionized. As the star contracts and the pressure increases, the layer changes its opacity and traps more heat, which drives the star back out. The driver exists only because of the overall structure of the star, which depends on its state of evolution, rendering all Cepheids grossly alike.

This similarity renders Delta and the rest of the Cepheids of great importance. In 1912, Henrietta Leavitt of Harvard University discovered that their luminosities are proportional to their periods. The more massive, and therefore larger, stars have the greatest surface areas and must be the brightest, but by their very sizes also take longer to pulsate. Consequently, a measure of a Cepheid's period gives its luminosity, which, in turn, (by comparison with apparent magnitude) gives its distance. Cepheids are so bright they are easily seen in other galaxies, which allows their distances to be found. Edwin Hubble's 1924 discovery of Cepheids in the nearby Andromeda galaxy led directly to our understanding of galaxies as separate entities and to the discovery of the expansion of the Universe. The examination of distant Cepheids was one of the Hubble Space Telescope's key projects, through which astronomers are stepping across the vastness of space to learn its character. And it all begins near home with Delta.

Delta Orionis, the right-hand star or Orion's shining belt, lies just below the celestial equator. Reddish BETELGEUSE marks Orion the Hunter's right shoulder (he is looking at you), Rigel his left foot. At the center of his sword, which dangles from his belt, lie the Orion Nebula and the Trapezium *(© Akira Fujii.)*

Residence:	ORION
Other name:	MINTAKA
Class:	CLOSE ECLIPSING DOUBLE, O9.5 + B0 GIANT.
Visual magnitude:	2.23 (COMBINED)
Distance:	900 LIGHT YEARS
Absolute visual magnitude:	–4.3 EACH
Significance:	THE "EQUATOR STAR" THAT HELPED US DISCOVER INTERSTELLAR GAS.

DELTA ORIONIS

As the Earth spins, the sky turns oppositely, pinioned to the north and south celestial poles. Between them, riding the sky from east to west, lies the celestial equator. From Earth's equator, the celestial equator passes overhead, from Earth's poles it lies on the horizon, and in between, it arches at an angle across the sky. In the planetarium, such points and circles can be lit by projection. At night we use Nature, the northern pole lit by famed **POLARIS**, the southern by obscure **SIGMA OCTANTIS**. Not as well-recognized is the "equator star," Mintaka, which lies only a quarter degree south of this celestial circle, making it a better marker than either of the pole stars.

As Delta Orionis, or Mintaka, rises and draws the heavenly equator, it leads Orion's belt across the sky. The name "Mintaka" comes from an Arabic phrase meaning "the belt of the Central One," who stands with **BETELGEUSE** as her right hand. Though Mintaka ranks seventh in Orion's brightness parade, Bayer, lettering the belt stars from west to east after naming the top three, made it Delta. Seventh in Orion, however, would be top in many other constellations, as Delta Orionis is still bright second magnitude.

A telescope reveals a companion, a sixth magnitude class B2 dwarf about one minute of arc away. At the star's distance of 900 light years, the separation is at least a quarter of a light year, and the orbital period is no less than 400,000 years. In between is a star something like our Sun, making the system a seeming triple. But now focus the power of the spectrograph on the bright component, on Mintaka proper. Instead of one set of absorption lines crossing the rainbow, we see two—one from an O9 dwarf, the other from a B0 giant, the pair far too close to be discriminated visually. We now have a quadruple star. Up close, the inseparable stars would appear similar to each other, each about 20 times heavier than the Sun and shining with some 70,000 times the solar luminosity. While Mintaka's O star is not as bright (either apparently or intrinsically) as the left-hand belt star Zeta Orionis (Alnitak, also meaning "the belt"), it is hotter and stands in well for all bright O stars.

Less than 0.2 AU apart (half the distance of Mercury from the Sun), the two stars orbit quickly. The Doppler effect makes the two sets of absorptions shift back and forth over a period of 5.73 days. The B giant just clips the edge of the smaller but brighter O star to produce periodic eclipses that cause a 20-percent dip in brightness. A century ago, the orbiting pair set the stage for a momentous discovery. While the bulk of the absorptions in Mintaka's spectrum are in continuous movement, in 1904, Johannes Hartmann found absorptions that did not shift. They could not have been produced by the orbiting pair but had to be made by intervening gas. Hartmann had discovered the "interstellar medium," the thin but pervasive gas that fills the spaces between the stars. We now know this medium to be enormously complex, to consist of dark dust and molecule-filled clouds nested within hierarchies of clumpy hotter gases. The dark clouds are the birthplaces of the stars.

The B giant is slightly evolved, having ceased nuclear burning in its core. Its mass is great enough to make an iron core that will collapse, exploding the star in a grand supernova. If the O star is not fatally damaged, it will explode as well, the two events leaving behind a pair of orbiting neutron stars. Long before Mintaka attains its final destiny, it will reach a different nexus in 2054, when the 26,000-year wobble of the Earth's axis will take the star across the celestial equator to mark it exactly.

If Deneb, shining in the tail of Cygnus the Swan, were as close as VEGA, it would shine as brightly as a crescent moon (which is here overexposed in a picture along with Comet Hale-Bopp). The Pleiades (with MEROPE) are up and to the right of the Moon, the head of Taurus (the Bull), with Aldebaran, up and to the left. *(J. B. Kaler.)*

Residence:	CYGNUS
Other name:	ALPHA CYGNI
Class:	A2 SUPERGIANT
Visual magnitude:	1.25
Distance:	2600 LIGHT YEARS
Absolute visual magnitude:	–8
Significance:	THE BRIGHTEST “A” STAR IN THE GALAXY.

DENEB

Flying south along the Milky Way, Cygnus, the celestial Swan, carries a variety of riches, among them the great star Deneb. Placed at the end of the Swan, its name aptly means "tail" in Arabic. Not confined to Cygnus, "Deneb" is used in some form for four other "tail" stars, among them **DENEB KAITOS**, which marks the tail of Cetus the Whale. Most classical constellations contain charming informal figures, or "asterisms," of their own, some more well-known than their parent constellations. Ursa Major (the Greater Bear) has the famed Big Dipper, while Ursa Minor has the Little Dipper, and Leo includes the "Sickle." Joining the crowd, Cygnus flies with the Northern Cross, which appears with Deneb at the top when the great bird is turned on its head. Asterisms can also extend over several formal constellations, and here, Deneb also shines as the eastern anchor of the Summer Triangle, one of a lovely trio of first magnitude stars that also includes Lyra's **VEGA** and Aquila's Altair.

All three of the Triangle's stars are class A, all white with temperatures in the neighborhood of 9000 K. As the sky's 19th-brightest star, near the end of the first magnitude list, Deneb is the trio's faintest. But only to the eye. Looking into the cosmos, it is easy to forget the third dimension. Although Deneb and its neighbor Vega are not all that different in apparent brightness, Deneb is 300 times farther away. To be first magnitude it must be hugely luminous, far more than Vega. The distance is uncertain, as the star is too far for direct parallax measures, but various estimates based on its spectrum and its association with other stars place it 1600 to 2600 light years away, although it is probably closer to the latter. The light we see departed during the time of ancient Greece. To appreciate Deneb's brilliance, move it to Vega's distance of "only" 25 light years. Assuming Deneb to be 2600 light years distant, the star becomes 10,000 times brighter and would shine in our sky at magnitude –8. It would be as bright as a crescent moon, would cast strong shadows upon the ground, and would wash out the faintest of the night's stars! Imagine the myths that would be told about it.

Deneb may be the most luminous white class A star in the whole Galaxy, shining nearly a quarter million times brighter than our Sun. For human life to survive its onslaught of radiation, Earth would have to orbit the star 400 times farther away than it does our own Sun, over ten times farther than Pluto. Were Deneb a main sequence star, one that, like Vega and the Sun, fuses hydrogen into helium in its core, it would be much hotter than it is. To be so luminous and still to have Vega's temperature, Deneb must be evolved and dying. It has quit core hydrogen fusion and has expanded to become a supergiant. It is indeed large, from its temperature and luminosity about 200 times bigger than the Sun, the size of the orbit of Earth.

Though the star is constant in its light, its spectrum is well-known to be slightly variable. Careful examination shows Deneb to be losing mass through a wind at a rate about 1000 times that of the Sun (through its "solar wind"). The star's great radiation is pushing matter outward and slowly whittling the star away. Deneb is probably fusing helium into carbon in its hot core, but we can only guess. Having begun its life as a star of some 20 solar masses, it will almost certainly explode sometime within the next couple of million years. Were it to do so now, it would cast your shadow with light created not long after Odysseus took his great journey through the ancient world.

29

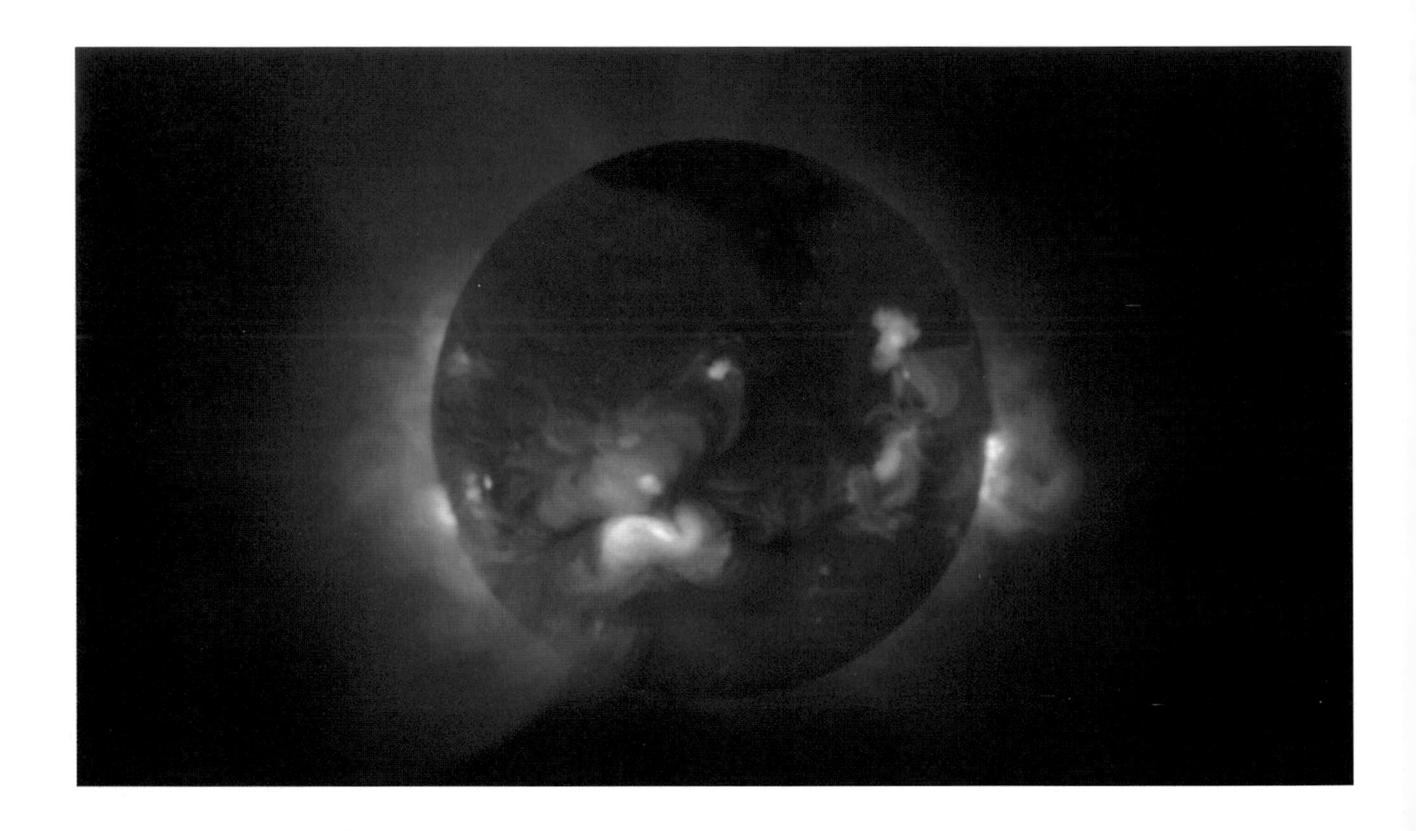

The Sun radiates X-rays from a surrounding corona that is heated by intense magnetic fields. Would X-ray-producing Deneb Kaitos look similar to the Sun? *(Yohkoh Observatory Soft X-Ray Telescope [SXT], prepared by the Lockheed Palo Alto Research Laboratory, the National Astronomical Observatory of Japan, and the University of Tokyo with the support of NASA and ISAS.)*

Residence:	CETUS
Other name:	BETA CETI
Class:	K0 ORANGE GIANT
Visual magnitude:	2.04
Distance:	96 LIGHT YEARS
Absolute visual magnitude:	–0.30
Significance:	A POWERFUL X-RAY SOURCE THAT SHOULD NOT BE.

DENEB KAITOS

No one would much notice another second magnitude star in Orion. Put one in dim Cetus and it shines like a searchlight. Lonely stars—like **ALPHARD** in Hydra—garner affection, and Deneb Kaitos certainly qualifies. Even so, the star receives little respect. Third magnitude Menkar (in Cetus's "jaw") is still the Alpha star, probably because Deneb Kaitos, (Arabic for "the sea monster's tail") is in the creature's nether-region, and so became Beta instead.

Outside of its prominence in an otherwise drab constellation mostly known as the residence of **MIRA**, Deneb Kaitos is a seemingly ordinary star. An orange giant just over the border into class K, it shines with a temperature of 4500 K. At a distance of 96 light years, it gleams with a luminosity 130 times solar which, while bright, is nothing special. The sky teems with such stars—dying swollen giants that have left the hydrogen-fusing main sequence behind. This one has about three times the mass and 20 times the diameter of the Sun which, compared to its brethren, is not that big.

But the star hides a remarkable enigma. Most cooler stars, including the Sun, emit X-rays as a result of magnetic fields generated by rotation and by convection in their outer layers. The magnetic fields are highly unstable. As they constantly form, change, and collapse, they transfer their energy upward to heat thin, vacuous layers above the stellar surfaces into the millions of degrees. The Sun's X-rays come from magnetically heated and confined loops of gas in the outer solar corona, the pearly layer that is visible only during a total eclipse.

Deneb Kaitos would have rotated rapidly when it was a hot main sequence star. When a body contracts, it spins faster, but when it expands, it slows down. As a star ages and grows to mature giant proportions, its rotation should come nearly to a halt. The change is gradual, and while the star is still becoming a giant, it may well generate magnetic fields and X-rays, but in older giants, their presence should be much reduced.

Lithium is a fragile element that is destroyed by nuclear reactions at relatively low temperatures. As it is continuously cycled downward by convection, we see less and less of it. Old stars have very little, if any. Lithium levels suggest that Deneb Kaitos should be a mature giant that is already fusing helium into carbon in its core. It should be, and is, rotating very slowly. Yet it is an amazingly strong stellar X-ray source, one of the most powerful in the sky, which suggests that it is just becoming a giant.

The enigma posed by Deneb Kaitos makes us realize that we do not understand stars as well as we might think. For example, consider well-studied **CAPELLA**. This close double is also a strong X-ray source. One of its components (Ab) is becoming a mature giant, the other (Aa) is already fusing helium. It is assumed that the X-rays are coming from the less-advanced star, yet Deneb Kaitos tells us that maybe it is the other one. Other studies show similar anomalies between lithium and magnetic activity. We are thus confronted with an important celestial mystery.

Why does it matter? The Earth lies in the Sun's magnetic environment and is powerfully influenced by it. To understand the origins of the solar fields, we had better have a theory that encompasses all stars. But we don't.

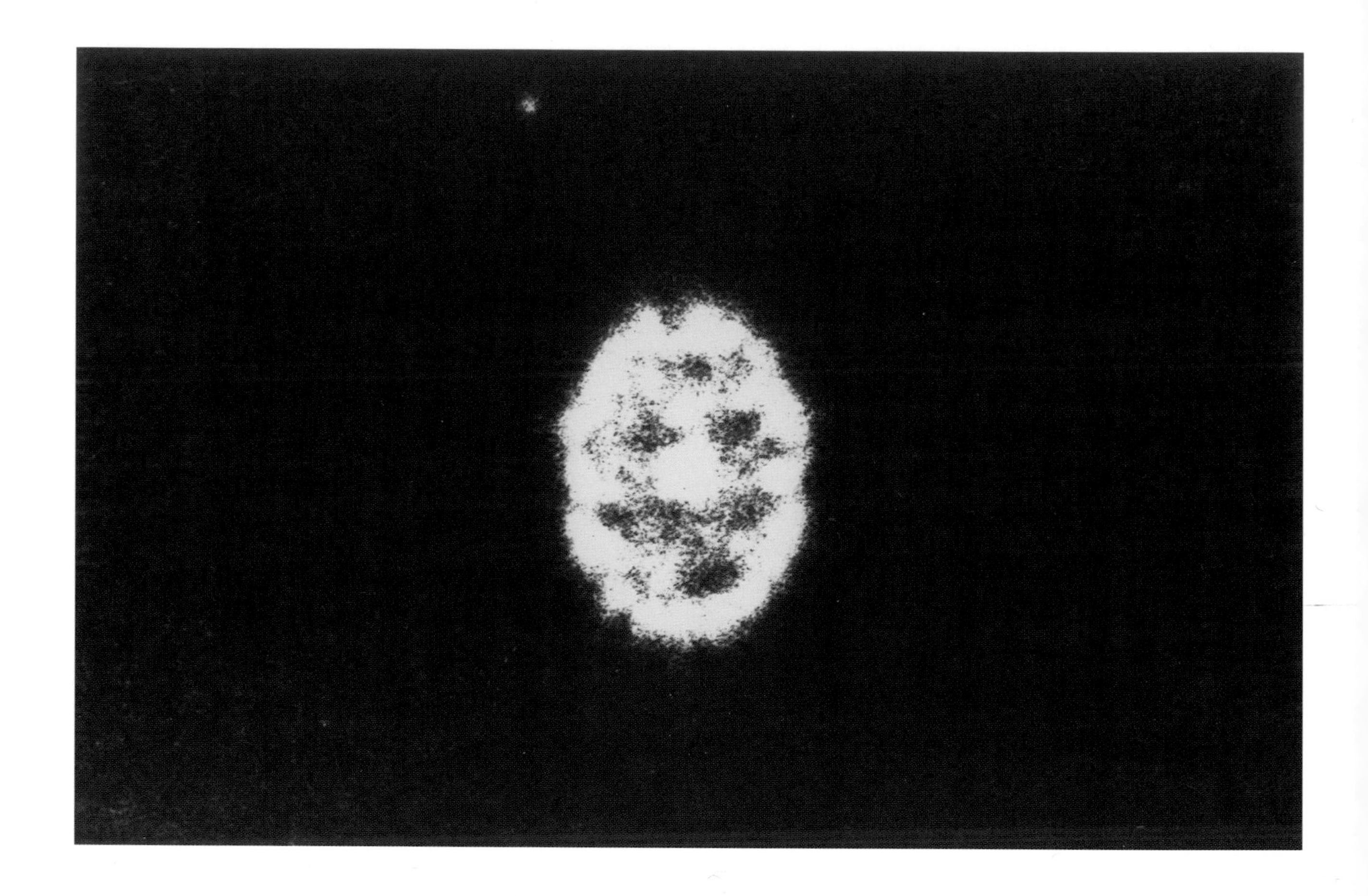

Forty years after outburst, the expanding shell of Nova Herculis 1934 (DQ Herculis at its center) had grown to a tenth of a light year across. *(R. E. Williams et al., University of Arizona photograph from the* Astrophysical Journal *224 [1978]: 171.)*

Residence:	HERCULES
Other name:	NOVA HERCULIS 1934
Class:	CLASSICAL DOUBLE-STAR NOVA
Visual magnitude:	14.4
Distance:	1600 LIGHT YEARS
Absolute visual magnitude:	5.3
Significance:	THE STAR THAT TOLD US HOW NOVAE WORK.

DQ HERCULIS

In 1934, a "new star" (from Latin, a "nova") of nearly first magnitude gave Hercules a new visage. A few weeks later, the Hero had returned to normal. About once every 20 years a similar nova graces the sky. Telescopic novae occur several times a year. "Nova," however, is a terrible misnomer. Once the event is over, old photographs show a faint star that must be the culprit, one that erupted to make itself known.

The term was first applied to **TYCHO'S STAR**, which was a much more violent "supernova," though of a different kind than the one that produced the pulsar in the **CRAB NEBULA**. Even in the 1930s, supernovae were thought to signal the destruction of entire stars. But after a period of years, fainter ordinary, or "classical," novae return to normal. A few years after an event, we also see a small expanding nebula, showing that some kind of explosion indeed took place. Suggestions ranged from stellar collisions to some kind of "atomic" bombs. In 1956, the star at the location of Nova Herculis 1934 told us the secret. It was found to be an eclipsing double like **ALGOL** and received the variable star name DQ Herculis. DQ, however, had a surprisingly short period of 4.65 hours, showing that the stars must be terribly close. Moreover, the combined light from the pair was found to vary with a period of only 71 seconds.

All novae are double stars. One component is a normal low-mass star, most commonly a class M star like **PROXIMA CENTAURI**, while the other one is an Earth-size white dwarf like **SIRIUS B**. The more massive component normally evolves first. As an expanding giant, its outer envelope interacts with the smaller star, such that friction brings the pair closer together. When the giant dies as a white dwarf, the two are so close that the dense white dwarf raises huge tides in the unevolved companion. The ordinary star's surface overflows the tidal boundary at which the combined gravity is effectively zero, and mass flows from the ordinary dwarf to the white dwarf. What we observe is not so much the stars themselves, but light from the heated flowing gas, which orbits the white dwarf before it crashes down.

White dwarfs are basically balls of carbon and oxygen. Whatever skin of hydrogen is on their surfaces is much too small and cool to fuse into helium. In a nova-to-be, however, the white dwarf becomes the companion's hydrogen dump. The white dwarf's gravity is so great that the base of the new hydrogen becomes compressed and heated. After 100,000 or so years, it explodes in a fusion bomb that blasts the surface of the white dwarf into space, causing a nova to erupt into our sky. From DQ's distance of 1600 light years, it must have put out the visual equivalence of nearly 100,000 Suns. The white dwarf is so dense, though, that its innards feel nothing.The system rather quickly returns to normal as it prepares to repeat the process many millennia into the future.

DQ not only "spilled the beans," it is fascinating in its own right. Normally, gas from a tidal companion first flows into a disk around the accreting star and then onto its equatorial surface. DQ's white dwarf, however, is highly magnetized and partially disrupts the flow. Most bizarre, the accreting gas hits off center and has spun up the white dwarf's rotation to a period of 71 seconds. Some kind of "beaming" then causes the 71-second light variations. Imagine standing on the white dwarf. Struggling against your weight of over 10 million pounds and through a blizzard of fresh hydrogen, you watch the starry sky rise and set once a minute, all the while hoping that your home will not explode beneath your feet!

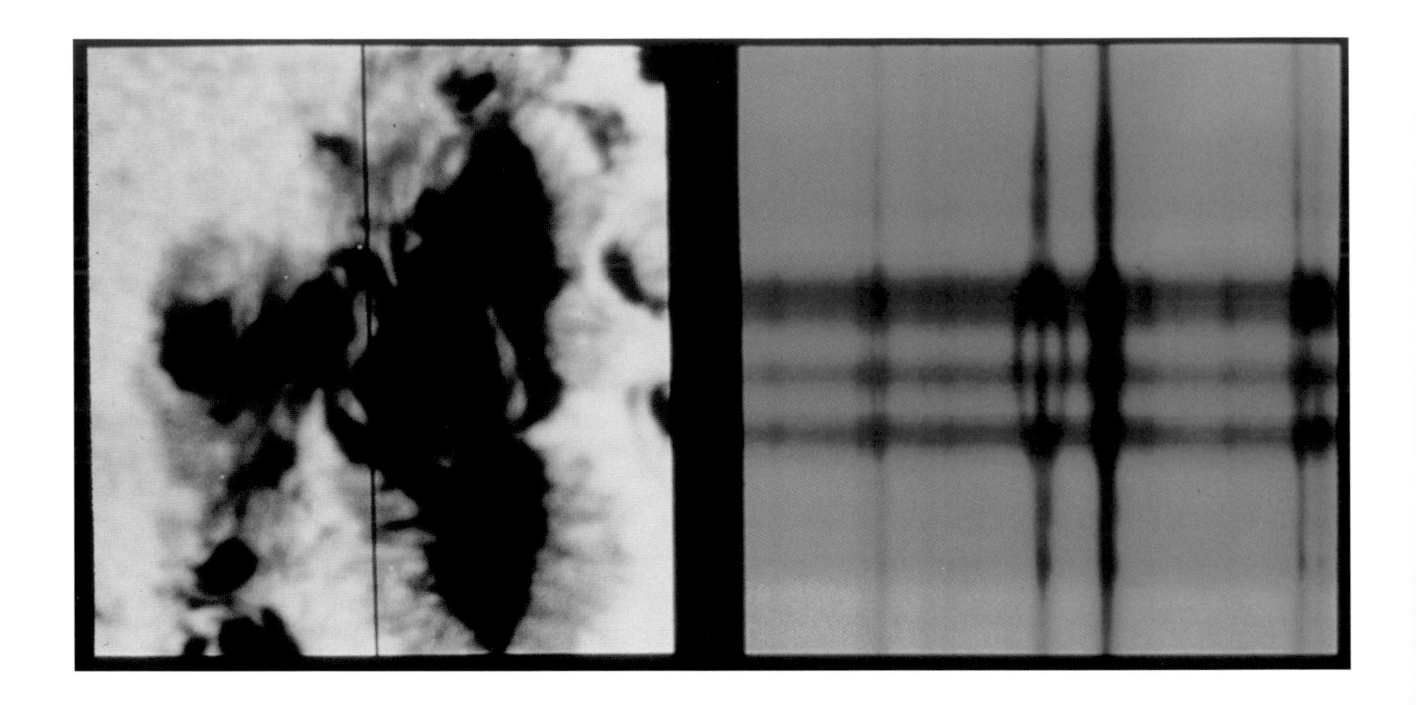

The powerful magnetic field of a sunspot (left), which is comparable to that in an MRI machine, causes dark solar spectrum lines to split into three parts (right), a process called the "Zeeman effect. "The field of EG 129, 50,000 times greater, makes hydrogen lines split across most of the visible spectrum. *(NOAO/AURA/NSF.)*

Residence:	DRACO
Other name:	LHS 3424 AND 17 OTHERS
Class:	MAGNETIC WHITE DWARF
Visual magnitude:	13.2
Distance:	42 LIGHT YEARS
Absolute visual magnitude:	12.7
Significance:	FIRST KNOWN, AND POWERFUL, MAGNETIC WHITE DWARF.

EG 129

Everything spins—the Earth once a day, the Sun once a month, pulsars in a second or less. Profoundly associated with rotation, and not all that well appreciated, are magnetic fields. Earth's is produced by the circulation of the iron in its molten core. Strong enough to make magnetized needles point north, the field is deeply rooted in the hearts of old navigators. The field also channels the solar wind—the flow of ionized gas from the Sun—to the poles and controls the northern lights; without it, the solar wind would impact directly on the Earth's atmosphere and impose a severe hazard to life. At the low end of the scale is the magnetic field of our Galaxy. With less than a millionth of the strength of our planet's field, it is generated by the Galactic 100-million-year rotation and is, in part, responsible for the formation of new stars like **T TAURI**. At the opposite end of the magnetic strength scale are magnetars like **SGR 1900+14**, with astounding fields 100 trillion times Earth's. White dwarfs, the ubiquitous deceased remains of ordinary stars of intermediate mass, are as a class slow rotators and would hardly be considered candidates for the magnetic hall of fame. But for still-mysterious reasons, about 2 percent of them are.

Astronomers have long known that magnetic fields distort spectrum lines (the "Zeeman effect"), the degree of splitting dependent on the field's strength. Absorptions from sunspots reveal fields several thousand times that of Earth, from peculiar stars like **COR CAROLI**, 5000 times. But nobody expected anything of white dwarfs; that is, not until the light from EG 129 (named after Olin Eggen and Jesse Greenstein), which displayed weird unidentified absorption lines, was discovered to be rather strongly "polarized." Polarization, in which light waves oscillate in a preferred direction, is also the result of a magnetic field. From the degree of polarization, EG 129 should have a field millions of times that of Earth. Though it was known to be peculiar among white dwarfs, no one particularly targeted EG 129; it was simply a part of a survey designed to look for possible magnetic fields, and out it popped.

Enter now the theoretical physicist, always waiting for just such an opportunity. Magnetic fields far more powerful than those in sunspots, and their effects, can be studied in the laboratory. But we can get nowhere near the field of EG 129. As a result, the theoreticians had to calculate the effect from what is known of the atom. The mysterious absorptions in EG 129 turned out to be those of simple hydrogen that had been not just split, but stretched and distorted all over the spectrum. The field's strength had to be more than 1 billion times that of Earth. The white dwarf had, in turn, become our laboratory for the test of theory!

We still do not know why such a small fraction of white dwarfs is magnetic, nor why they are magnetic in the first place. Nor have we found their predecessors among the central stars of planetary nebulae. But the magnetic white dwarfs are more massive than most, and they may be the leavings of peculiar class A and B stars like Cor Caroli. Or maybe not. The tiny stars still challenge.

Dust shells now nearly one light year across, resulting from thousands of years of stellar mass loss, surround the central star of the Egg Nebula. The rapidly evolving star is quite hidden by the thick bar of dust at the center. *(Hubble Space Telescope, R. Sahai and J. Trauger [JPL], the WFPC2 Science Team, and NASA.)*

Residence:	CYGNUS
Other name:	AFCRL 2688
Class:	PROTOPLANETARY NEBULA WITH CLASS F SUPERGIANT
Visual magnitude:	APPARENT MAGNITUDE OF STAR: INVISIBLE
Distance:	4000 LIGHT YEARS
Absolute visual magnitude:	VISUAL MAGNITUDE OF STAR:–6?
Significance:	A RAPIDLY EVOLVING STAR THAT IS PRODUCING A PLANETARY NEBULA.

EGG NEBULA

An invisible star? How do we know it is there if we cannot see it? The human eye, as good as it is, has a very limited wavelength range. We cannot get a clear understanding of the Universe unless we can observe it in all wavelength bands, from gamma rays to radio. The infrared part of the spectrum is crucial for examining cool objects in space that do not radiate in the optical spectrum. However, infrared radiation penetrates the Earth's atmosphere only in certain narrow bands, so a complete look requires that we go into space. (Atmospheric trapping of infrared light radiated by the cool Earth, the greenhouse effect, helps keep us warm.) In the 1970s, the US Air Force Cambridge Research Laboratories carried out a high-altitude infrared survey of the sky with rockets, their instruments getting a mere glimpse before returning to Earth. One source—AFCRL 2688—stood out brightly at the far-infrared wavelengths of 10 and 20 micrometers. Visible-light photography showed a small elongated object that was dubbed the "Egg Nebula," certainly appropriate for its residence in Cygnus, the celestial Swan.

Deeper examination revealed a double nebula, the components separated by only 7 seconds of arc with the infrared source lying between them (the "Broken Egg?"). Curiously, the nebular parts had identical spectra, those of a class F supergiant star. Moreover, the light from the two lobes was seen to be highly polarized, that is, oscillating in one direction. Polarized light—easily detected with polarizing sunglasses—is commonly produced in Nature when light is reflected from a surface.

Astronomers quickly realized that the little nebulae must be clouds of dust that were reflecting the light of a brilliant F star embedded in a disk-like cloud between them—a disk so thick that visible starlight could not get out. The star heats the central cloud, which, in turn, radiates its light in the far infrared. Starlight directly escapes through the poles of the disk and is reflected by the surrounding thinner nebulae to produce the "Egg."

Better images at a variety of wavelengths, culminating with two made by the Hubble Space Telescope, clearly confirm this picture. Hubble reveals a spectacular sight, the near-twin lobes now seen as doubled searchlight beams shining through murky, dusty clouds, each embedded with dozens of circular rings. Running through the center, perpendicular to the lobes, is the black opaque disk that hides the brilliant star within. Though 4000 light years away, the star would be visible to the naked eye were it not for its shroud.

Even in the 1970s, the object looked suspiciously like a star that was changing from a giant star into a planetary nebula like **NGC 6543** or **NGC 7027**. When the helium-burning that powers a red (class K or M) giant star stops, the star expands to huge proportions and eventually begins to pulsate to create a cool, long-period variable star like **MIRA**. At the same time, the star loses most of the huge distended envelope that surrounds its old nuclear burning core through a powerful wind that is expelled mostly at the star's equator. Dust and molecules then form within the cooling wind. The near-exposure of the fiery core makes the star's new surface hotter and hotter, the star now passing through warmer classes, from M to K to G to F and so on. We see the Egg in just such a transition. The rings in the Hubble picture show that, for reasons not understood, mass loss is episodic. Eventually the star will become so hot that it will ionize the surrounding gas and partially dissipate the dust. The Egg will be fried, and a new planetary nebula will be born.

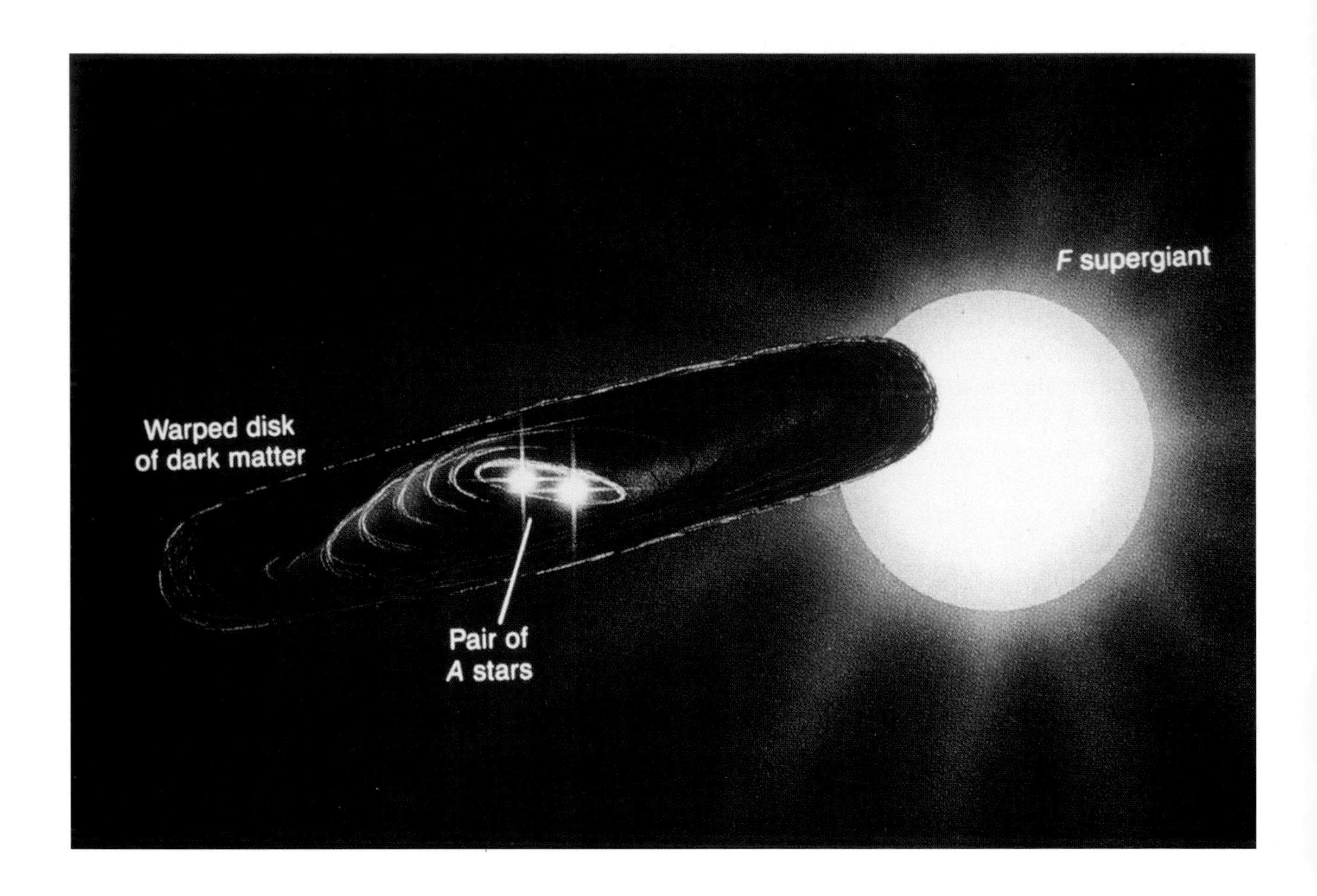

Epsilon Aurigae (see also #18), a supergiant, is periodically eclipsed by a huge cloud of dust that seems to harbor a pair of B (or perhaps A) stars. *(Courtesy of Sky and Telescope [January 1988]: 15.)*

Residence:	AURIGA
Other name:	ALMAAZ
Class:	F0 SUPERGIANT
Visual magnitude:	3.04
Distance:	3300 LIGHT YEARS
Absolute visual magnitude:	–7.1
Significance:	A SUPERGIANT ECLIPSED BY SOMETHING EVEN BIGGER.

EPSILON AURIGAE

Almost lost among the northern winter's brilliant stars and of lesser fame than his mighty neighbor Orion, Auriga the Charioteer is best-known as the home of the first magnitude giant double star **CAPELLA**, the "She-Goat." A lovely pentagon-shaped figure connected to Taurus, the Charioteer is oddly depicted as carrying said goat (a goat chauffeured by chariot?) as well as her three Kids, represented by a flat triangle of lesser stars to the southwest. Equally charming, the Kids, Capella, and two of the constellation's eastern stars, seem to make a giant celestial arrow that points west to Perseus.

While Capella dominates, the third magnitude Kids (Epsilon, Zeta, and Eta Aurigae—Bayer missed the boat in not naming them Zeta, Eta, and Theta) are much more entertaining, as perhaps expected of lively little ones. The brightest and faintest of the Kids, Zeta and Epsilon, are variable, while the middle one, Eta, serves nicely as a steady comparison against which to follow the meanderings of the others. Both variables are eclipsers.

Zeta, the fainter of the two, is a magnificent double. Every 2.6 years, a small B5 dwarf 400 times more luminous than the Sun disappears behind an enormous K giant, the orange K star so big—about the size of the Earth's orbit—that the total eclipse takes 38 days. The event is difficult for the naked eye to make out, because the giant is five times more luminous than the dwarf and thus dominates the system. During an eclipse, the larger star's total visual light drops by only a couple tenths of a magnitude when the little B star disappears. The B star radiates very strongly in the invisible ultraviolet, where the eclipse appears much greater.

It is good first to be impressed by Zeta, as it sets us up for absolute amazement when confronted by Epsilon (originally not one of the Kids, but added later). There is no problem seeing the eclipse with the naked eye, providing one lives long enough. The star we see is an F supergiant twice the Earth's orbital diameter in size, radiating some 60,000 times the light of the Sun. Every 27 years, something obscures up to half of it for almost two years, the visual light falling in half, by 0.8 magnitudes, with the period of minimum light lasting for 380 days.

Epsilon Aurigae is the only known supergiant to be eclipsed in this way. What could be big enough to block half the light of a supergiant? The orbital characteristics of the F star indicate an orbiting body of large stellar mass. The eclipse takes so long that, if the eclipsing body were actually a star, it would be big enough to obscure the supergiant completely. Yet during the eclipse, the supergiant remains visible. Moreover, nothing else is visually evident.

Infrared radiation from a cool "something" glowing at about 500 K helps give the mystery away. Much too cool to be a star, the "something" appears to be a vast cloud of orbiting dust. But no distended cloud, however massive, could have any permanent integrity. The most likely—indeed the only—explanation is that the dark cloud contains a visually hidden orbiting star. Given the estimated mass, the star would have to be an O dwarf. But an O dwarf should be visible. To make the right mass, the cloud likely contains a pair of cooler and less luminous B dwarfs that orbit each other, this double and its shroud all orbiting and obscuring the supergiant. How the dust is generated by the double is quite unknown, and as far as we know the system is one-of-a-kind.

Perhaps the next eclipse will tell us more. Prepare now: it is scheduled to take place in 2009.

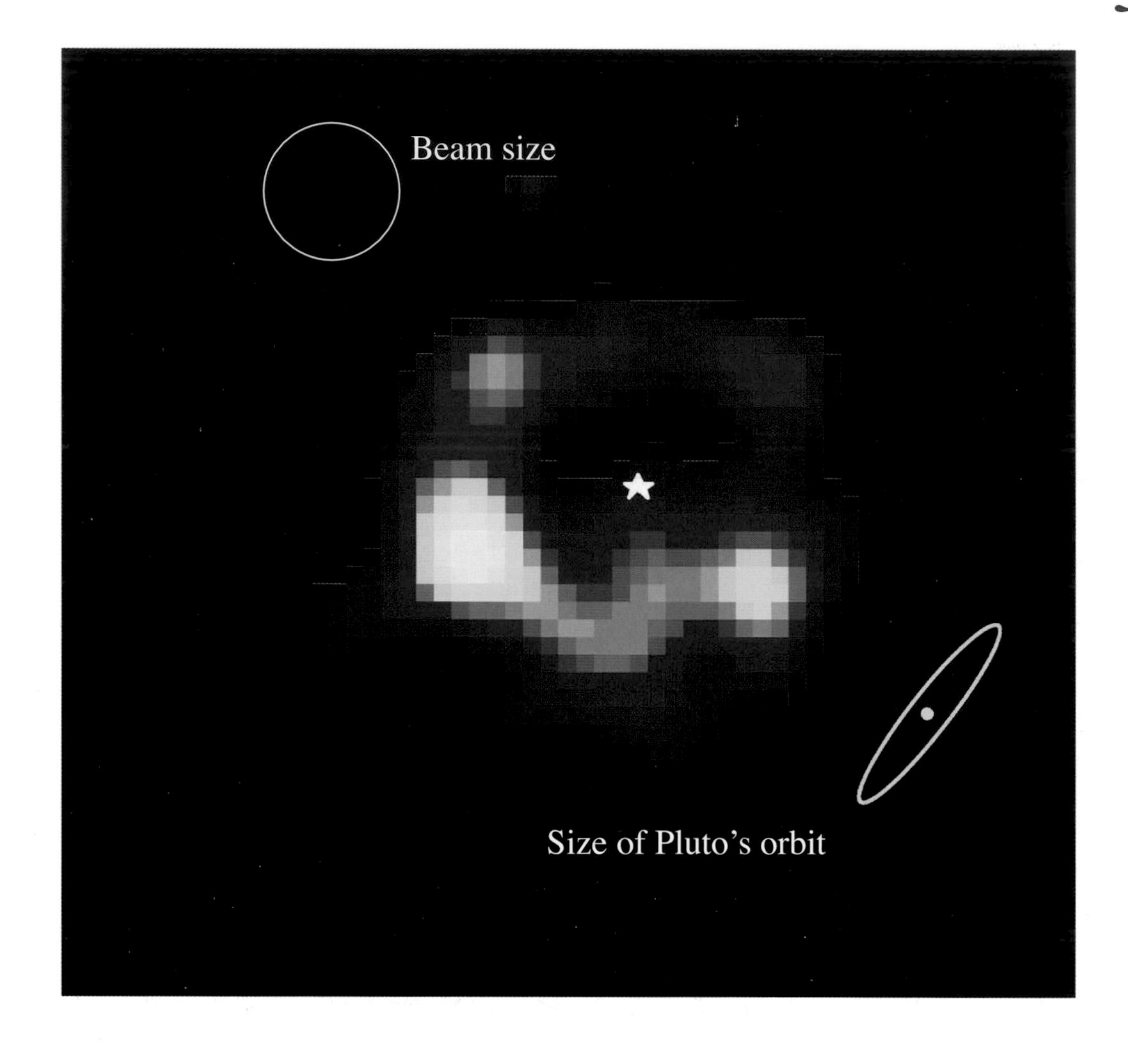

A dusty ring, seen by its long-wave infrared emission, surrounds Epsilon Eridani, a star known to have a planet. The hole near the center is about the size of the orbit of Pluto (ellipse at lower right), suggesting the existence of a planetary system. The ring might be analogous to our System's Kuiper Belt of comets. *(J. S. Greaves et al., James Clerk Maxwell Telescope, Joint Astronomy Centre, United Kingdom PPARC, Netherlands OPP, and NRC Canada.)*

Residence:	ERIDANUS
Other name:	18 ERIDANI
Class:	K2 DWARF
Visual magnitude:	3.73
Distance:	10.5 LIGHT YEARS
Absolute visual magnitude:	6.19
Significance:	AN ORIGINAL SETI STAR AND THE CLOSEST WITH A CONFIRMED PLANET.

EPSILON ERIDANI

SETI—the Search for ExtraTerrestrial Intelligence, the hunt for extrasolar civilizations. Are we alone? To have "ETI," you must have planets. The search goes forward on both planes. Look for planets and then speculate on life, or look for "intelligent" radio signals, their existence taken as proof of planets: Epsilon Eridani serves both camps.

Epsilon Eridani is a fourth magnitude class K2 ordinary dwarf. At 10.5 light years away, it is the ninth-closest star to Earth (counting doubles and multiples as units), and the third-closest visible to the naked eye. Its temperature of 5100 K and low luminosity (about 35 percent that of the Sun) tell of a lower mass, just under three-quarters solar, all befitting a star farther down the main sequence than the Sun. Of much greater interest, Epsilon Eri displays great solar-like activity. Active regions that contain large starspots ("visible" not directly, but by spectroscopy) rotate in and out of view, yielding a rotation period of only 11.1 days, less than half the solar period. In turn, the faster rotation is responsible for the increased magnetic activity. Since stellar magnetism coupled with an outflowing wind can slow a star's rotation, the rapid spin suggests a younger age, only about 15 percent the solar age of 4.5 billion years. Epsilon Eridani even has an activity cycle of 5 years, about half the length of the famed 11-year solar cycle.

The star's proximity and similarity with the Sun suggested to Frank Drake that it was a good candidate for "listening" for communication signals perhaps produced by an advanced civilization. So, in 1960, he and the others of Project Ozma took a radio "look" at both Epsilon Eridani and similar Tau Ceti (a G8 dwarf 11.9 light years away). A false signal from Epsilon sparked instant interest. Nothing has been heard since.

Could there still be a planet? More false signals. A wobble in the position of the star "discovered" in 1974 was not really there. Later, tantalizing velocity shifts again suggested a planet that might be moving the star back and forth, as was proven to be the case for **51 PEGASI**, the star with the first discovered and confirmed planet. Finally, as the century turned, there was no question. Astronomers had confirmed that Epsilon Eri was the closest star known to have a planet in orbit about it.

From the minute Doppler shifts recorded from the stellar spectrum, we find the planetary characteristics. The body is not one that would support ETI. With an orbital radius of 3.3 AU, it is among the farthest of the extrasolar planets away from its parent star. A rather high orbital eccentricity takes the body from as close as 1.3 AU (about the distance of Mars from the Sun) to as far as 5.3 (near to what would be Jupiter's orbit). At such a distance, the orbital period is long, about 7.1 years, making it a difficult "find." Hardly a planet like Earth, its mass is quite similar to that of Jupiter; we would expect the structure to be similar as well, noxious atmosphere and all.

However, where there is "Jupiter," there may be "Earth." Even "Earths." Infrared radiation from the system reveals a dust ring satisfyingly similar to the one around our own Sun, which contains the Kuiper Belt of comets beyond Neptune. Its asymmetry and a hole in the middle suggests that planets have formed through dust accumulation, just like our own System. Eventually, we may see Epsilon's "Jupiter" and maybe even find its "Earths," though the high eccentricity of the large planet's orbit may well make smaller inner planets gravitationally impossible. And so far no ETI. SETI moves on, for now passing Epsilon Eridani by.

The naked-eye double Epsilon Lyrae lies just down and to the left of brilliant VEGA. Each of the stars is again broken in two. The parallelogram down and to the right of Vega holds BETA LYRAE at its upper right corner. *(J. B. Kaler.)*

Residence:	LYRA
Other name:	THE DOUBLE-DOUBLE IN LYRA
Class:	FOUR A DWARFS
Visual magnitude:	3.88
Distance:	160 LIGHT YEARS
Absolute visual magnitude:	1.7 TO 2.6
Significance:	A NAKED-EYE DOUBLE AND THE CLASSIC DOUBLE-DOUBLE STAR.

EPSILON LYRAE

William Herschel must have been delighted in 1779 when he turned his telescope to a fourth magnitude star just northeast of **VEGA**, one that helps make the exquisite constellation Lyra (the Harp). Long known to be double, Epsilon Lyrae is a supreme test of naked-eye vision, its two components separated by 3.5 minutes of arc. With high-enough telescopic power, each again breaks into a double 80 times closer, each subset now just 2.5 seconds of arc apart. Here, any amateur with a decent rig can see four nearly identical stars. Together, they make the famed "Double-Double in Lyra." Excluding Vega, Alpha Lyrae, the five stars that make the celestial Harp are roughly the same brightness, and their Greek letters are called out in a spiral order, the Double-Double getting "Epsilon." By astronomical tradition, the western-most of the naked-eye pair is called Epsilon-1 (ϵ^1), the other ϵ^2.

The stars within each pair are so close together that their properties have been rather difficult to determine. Apparent magnitudes average 5.5, classes A5, while temperatures hover near 8000 K and luminosities about 12 solar. A closer look reveals subtle differences, especially between the two components of Epsilon-1, as one is noticeably fainter than the other. The four are ordinary white hydrogen-fusing main sequence dwarfs with masses and radii about double those of the Sun. All spin with at least 100 times the solar rotation speed. All at the same distance of 160 light years, the stars of the close pairs each orbit the other, while at the same time each of the pairs are in mutual orbit. The members of each of the close pairs are separated by about 140 AU (3.5 times Pluto's distance from the Sun), and it takes them about 1000 years to make the journey. The orbital period of each double about the other is vastly longer. The pairs are separated by at least 10,000 AU (0.16 light years), implying a period of at least half a million years (making orbital observations rather obviously impossible).

While Epsilon is surely the most famed, such double-doubles are not that rare. **MIZAR** in Ursa Major is similar except that there the wide pair is telescopic and each of these can be split only with highly sophisticated techniques. Castor, Alpha Geminorum, is another fine example. Such doubling is hardly accidental. As stars are born by condensation from dusty interstellar gases, they spin faster, and when they are rotating too quickly, they divide to make a double star. The first set of components of Epsilon Lyrae and its kin simply divided yet again.

While many class A stars (like **BETA PICTORIS**) reveal dusty disks in which planets might have formed, there is no evidence that Epsilon Lyrae's stars have either; duplicity may rule against planetary formation (we have no idea). But try to visualize the sight if there were one. An "Earth" would have to orbit one of the stars at about Jupiter's distance from the Sun for us to survive. The other of the pair would shine as a second "Sun" with a radiance of 500 full moons. Off in the distance, we would see two brilliant points under 1 degree apart, each of which glows with the light of a quarter Moon. Imagine too the mythologies that our planet's inhabitants might have invented, not to mention the songs they might sing.

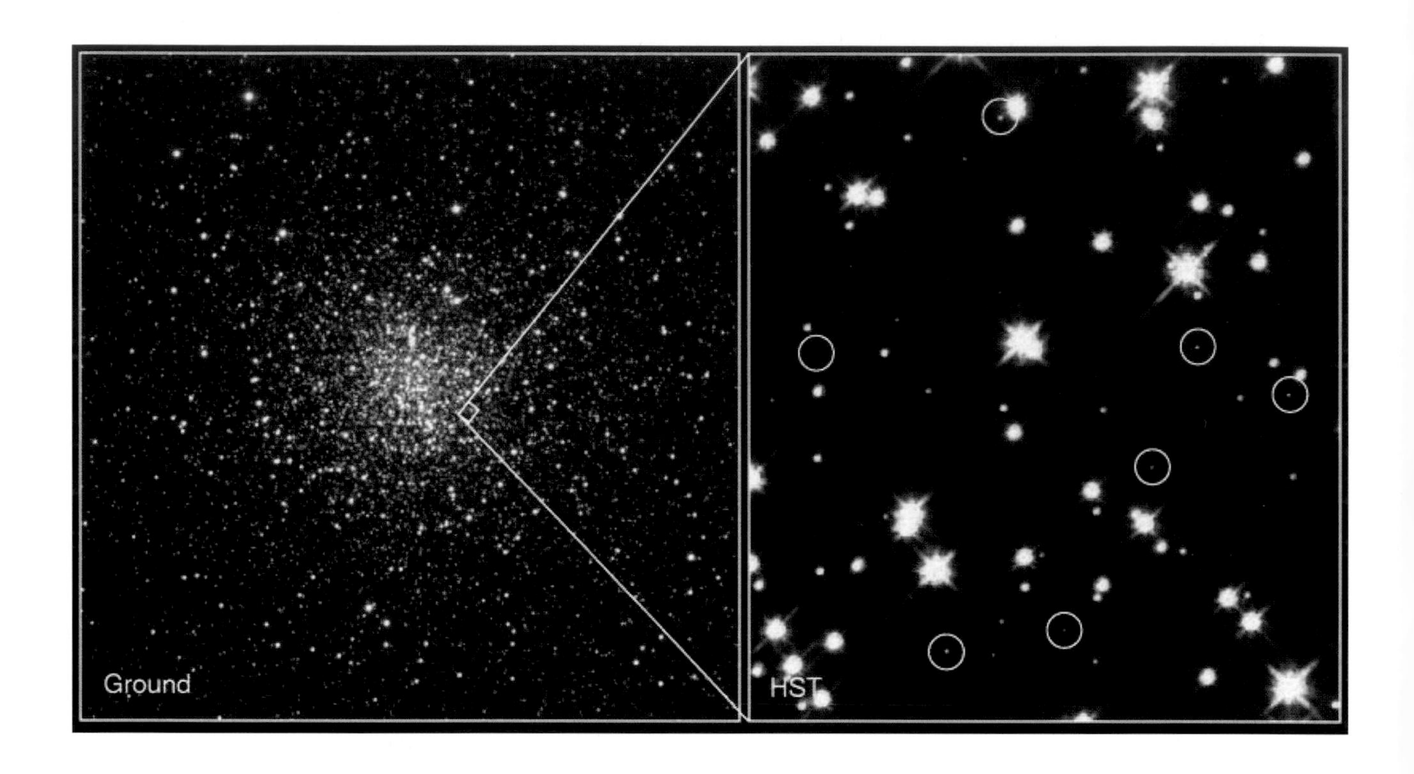

Ancient dim white dwarfs flock the center of the globular cluster Messier 4. *(Hubble Space Telescope, H. Richer [University of British Columbia], H. Bond [STScI], and NASA.)*

Residence:	HYDRA
Other name:	NONE
Class:	WHITE DWARF
Visual magnitude:	20.5
Distance:	140 LIGHT YEARS
Absolute visual magnitude:	17.4
Significance:	THE FAINTEST KNOWN WHITE DWARF.

ESO 439–26

The strange nature of white dwarfs is best illustrated by the nearest, **SIRIUS B**. Why would anyone care about the faintest of the breed? Because as a record-holder, it stands in for the whole batch of faint white dwarfs that tell us about our Galaxy and its age. And it has its own story to tell, too.

White dwarfs are the ancient nuclear-burning cores of main sequence stars that have lost all their outer envelopes (along the way producing things like the **EGG NEBULA** and planetary nebulae such as **NGC 6543**). They come from stars that initially ranged from 0.8 to around 10 solar masses. At the low end, the stars make white dwarfs of around half a solar mass. At the high end, they are at the "Chandrasekhar Limit" of 1.4 solar masses, beyond which the cores collapse and explode to make supernovae. Small (the size of Earth) and dense (1 ton in a sugar cube), they begin their white-dwarf lives above 100,000 K, as their ejecta dissipate into space. Their subsequent fate is to cool and dim forever.

For all their seemingly strange properties, white dwarfs (like astronomers) are relatively simple things, so theoreticians have little trouble calculating their cooling and dimming rates. Higher-mass stars burn out relatively quickly. From the ages of the ancient globular clusters, which are determined by the most massive stars that remain within them, we know our Galaxy is around 13 billion years old. The cooling and consequent dimming of white dwarfs is so slow that even after all that time, the oldest has yet to chill to the point where it cannot be seen. Consequently, all we need to do is look for the coolest and faintest white dwarfs we can find, apply theory, and we have another way of measuring the Galaxy's age. Globular clusters are found only in a vast halo that encompasses the Galaxy's thick stellar disk. White dwarfs are, by their natures, very faint, and they cannot be seen at great distances. We are, therefore, far from having a good sample of them within the distant halo. But we do see the little stars all over the disk. Here, they quit at a magnitude of 16. The theory of their dimming shows that the oldest are, like the disk, roughly 10 billion years old.

While its chief attribute is in calling attention to dim white dwarfs, ESO 439–26 is still singular. Holding the faintness record with an absolute visual magnitude of 17.4, it is actually dimmer than the standard white-dwarf cutoff. Some 140 light years away, it is less than one hundredth the apparent brightness of Pluto. To be visible to the naked eye, it would have to be within 0.2 light years, only 5 percent the distance to **ALPHA CENTAURI**. We might expect the faintest white dwarfs to be coolest, but while ESO 439–26's luminosity may be lowest, its temperature is not. At 4560 K, it is warmer than the current record of 3500 K for a somewhat brighter star called WD 0346+246. (At these temperatures, "white" dwarfs are reddish, showing the danger in naming classes of objects before they are understood.)

Why the anomaly? Normal main sequence stars become quickly brighter and larger as their masses increase; this behavior is the result of higher internal heat and compression as well as greater rates of nuclear fusion. White dwarfs, however, are effectively dead. Higher masses lead to higher gravitational fields that make them smaller. With less surface area for the same temperature, higher-mass white dwarfs are fainter. While most have masses around 0.6 solar, to explain its anomalous dimness, ESO 439–26 must have a mass even greater than that of Sirius B. Dim it might be now, but it began life as a brilliant, hot class B dwarf. How the mighty do fall.

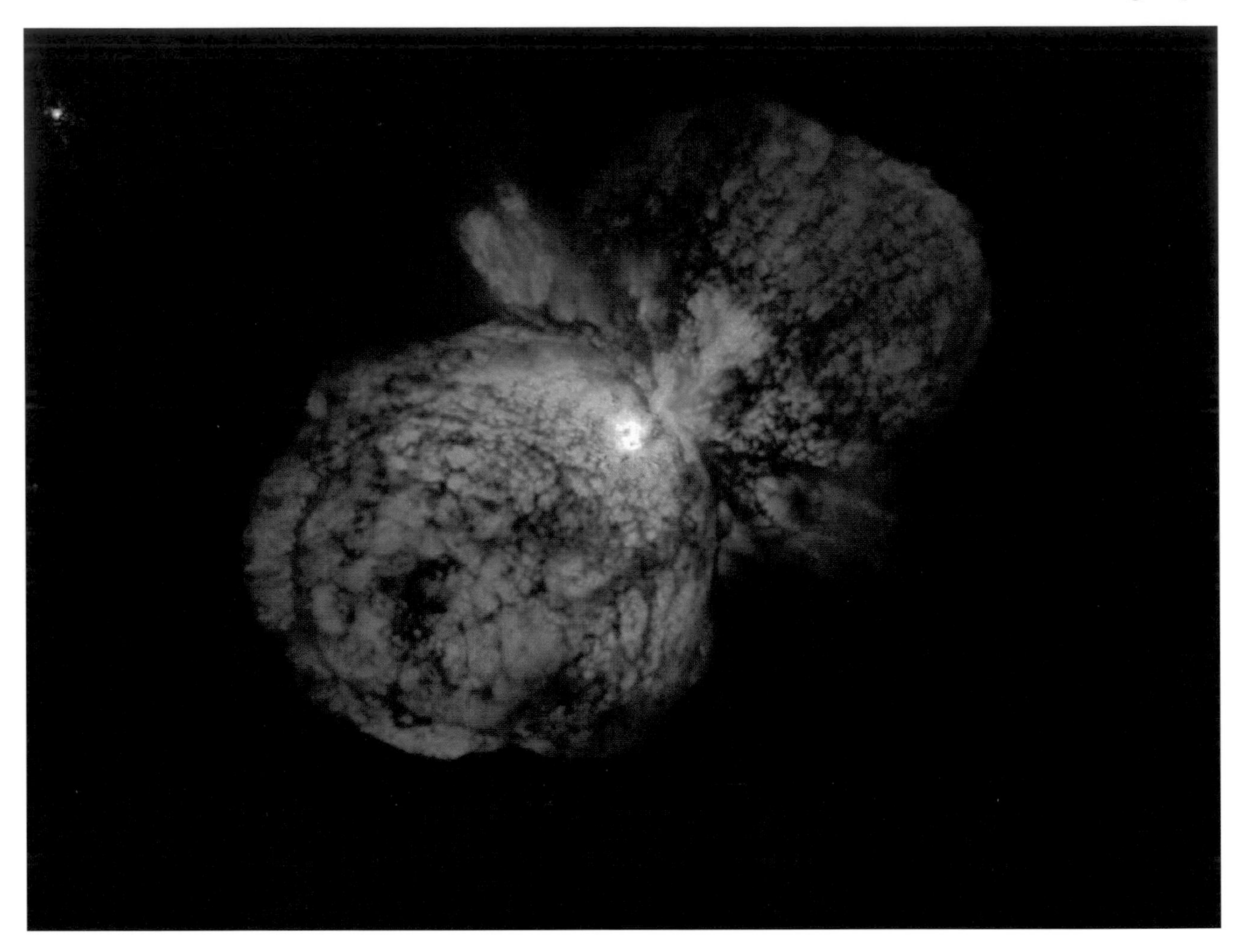

The supermassive star Eta Carinae (see also #50), buried at the center, belched two expanding dusty lobes during its mid-nineteenth century outburst. Now over half a light year across, the lobes are centered on a dusty disk ejected over 50 years later. *(Hubble Space Telescope, J. Morse [University of Colorado], STScI, and NASA.)*

Residence:	CARINA
Other name:	NOVA CARINAE 1843
Class:	B0 HYPERGIANT
Visual magnitude:	5
Distance:	8000 LIGHT YEARS
Absolute visual magnitude:	–10
Significance:	NEAR THE UPPER MASS LIMIT AND ABOUT TO BLOW.

ETA CARINAE

Eta Carinae needs no dramatic buildup: almost 5 million times brighter than the Sun and 500 billion times brighter than the faintest white dwarf, **ESO 439–26**, it is one of the most massive stars in the Galaxy. Buried in a dense cloud of its own making (the "Homunculus," from its vague resemblance to a human figure), it is also one of the most mysterious.

Among the most distant stars visible to the naked eye, 8000 light years away, Eta now shines at fifth apparent magnitude. It has not always appeared so modest. During the 17th and 18th centuries, it varied between magnitudes 2 and 4. Around 1840, the star erupted. Reaching magnitude –1, it glowed brighter than nearby **CANOPUS** and almost rivaled **SIRIUS**. Within 30 years, Eta had faded to eighth magnitude invisibility. Burping back to sixth magnitude in 1890, it faded again and has only recently resurfaced to the eye.

Imaging with Hubble shows the Homunculus to be made of two stunning dusty lobes nearly a light year across and bisected by a ragged waistband. Moving outward at a speed of nearly 400 kilometers per second, the lobes are a huge bipolar flow, twin rivers of matter moving in opposite directions from the star. They were produced by the Great Eruption of 1840 when Eta Carinae belched over a solar mass of gas. The waistband is a faster flow that came from the 1890 eruption. The star's luminosity suggests a mass around 100 times solar, so as great as the eruption was, the ejecta were hardly missed. Much of the gas produced in the eruption condensed to dust, which shrouded the star and caused the great fading. The current brightening is probably due to the thinning of the dust as the Homunculus expands, and the star begins to reveal itself, slowly returning to its old third-or-so magnitude level. Inside the nebula lies a blue class B supergiant, one so luminous that a new category of "hypergiant" must be applied. Yet even among hypergiants, Eta is rare. Its variations have proved it to be one of a handful of "luminous blue variables."

Or is Eta Carinae two stars? Changes in the nebular spectrum lines reveal a 5.6-year period that implies a binary nature. If so, it must be one of the most massive doubles in the Galaxy. Intriguing evidence arrives from the Homunculus, which is rich in nitrogen and carbon, but low in oxygen. Such a composition is expected from a star that has run on the "carbon cycle," a mode of hydrogen fusion in which carbon is used as a nuclear catalyst. Yet the actual star as seen through the dust in infrared radiation is not so enriched. The Great Eruption may have been produced not by the visible star, but by the companion, a star that is now almost completely hidden by dust and has stripped itself down through a great wind to reveal its nuclear fused innards. The companion, which must once have been the more massive, is now the less massive (the **ALGOL** paradox revisited). Current estimates suggest that the "primary" star (once the secondary) still has around 80 solar masses, while the secondary (once the primary) has perhaps 60. The two may have been born with 200 solar masses! Winds from the rapidly evolving stars collide, heat, and radiate X-rays.

Whether one star or two, Eta Carinae will explode. If the star is indeed double, the invisible secondary will go first (as it is the farthest along its evolutionary path). The wisdom is that Eta Car will produce a hypernova, a grander version of the supernova in which the stellar core collapses to form a black hole. In the process, the star may create a burst of gamma rays that will be seen clear across the visible Universe. Maybe even two of them.

A complex planetary nebula surrounds the weird star FG Sagittae. *(G. H. Herbig, University of Hawaii Observatory.)*

Residence:	SAGITTA
Other name:	HENIZE 1–5
Class:	PLANETARY NEBULA CENTRAL STAR
Visual magnitude:	9
Distance:	AN UNCERTAIN 8000 LIGHT YEARS
Absolute visual magnitude:	AN UNCERTAIN –3
Significance:	A "ROSETTA STAR."

FG SAGITTAE

Most stars do not change over a human lifetime. Though they may vary, they do not fundamentally alter their natures. To learn how stars age, we use theory to string the various kinds together. What a marvel, then, to see a star transform itself as we watch.

FG Sagittae is centered in a faint planetary nebula called Henize 1–5. When first recorded in 1896, it had a "blue" magnitude (observed with a blue-light-sensitive photographic plate) of 13.6. Brightening steadily, by 1969 it was up by 4 magnitudes, and by 1991, the standard visual magnitude stood at 9.0. The brightness change was a bit of a sham. The actual total luminosity never changed much. Instead, the star was cooling, changing from a class B4 supergiant in 1955 (with a temperature near 15,000 K) to class G2 (6500 K), where it is today. When hot, it radiated much of its light in the ultraviolet, and as it cooled, the radiation just shifted into the part of the spectrum we can see.

Such behavior alone would have produced little more than modest excitement. What galvanized astronomers were the compositional changes. In 1960, the chemical composition of the star's surface gases were similar to those found in the Sun. Then the "rare-earth elements" arrived. Along with lead and neighboring elements, the rare earths (which lie chemically between barium and hafnium and include such oddities as cerium, praseodymium, and samarium) grew to become thousands of times more abundant than they are in the Sun. Most intriguing, these are "s-process" elements, built through the slow, steady capture of neutrons onto lighter elements. In addition, FG Sagittae became highly enriched in carbon. Some momentous event must have taken place.

As the central star of a planetary nebula, FG Sagittae was once a huge giant like **MIRA** that cast off its outer envelope to reveal its core (now seen as the star), creating the nebula in the process. "Miras" and their non-varying cousins (the collection of "second-ascent" giant branch stars) have dead carbon-oxygen cores surrounded by two shells. The inner shell is fusing helium into carbon, while the outer is burning hydrogen into helium. As such stars age (so goes theory), the shells alternately turn on and off. The outer goes rather quietly, but the inner erupts violently in an internal helium flash. Elements created within the burning maelstrom find their way to the top on the coattails of convection, and the stars change their kinds, some becoming carbon stars like **R LEPORIS**.

But the central stars of planetary nebulae are turning into white dwarfs! They are hardly giants, and nuclear fusion has all but shut down. Deep within, however, lurks the helium-rich layer that surrounds the carbon core. In a small number of cases, the conditions (whatever they are) are just right for one last gasp, one last helium flash. The immense energy released expands the star, and what was once a near-white-dwarf is transformed back into a giant, perhaps even into a supergiant. At the same time, a wealth of new atoms created in the helium-fusing zone make their appearances.

In 1992, FG Sagittae suddenly faded by nearly five magnitudes, becoming only one-one hundredth as bright. Recovering quickly to tenth magnitude, it dropped again, with the up and down behavior continuing to this day. FG Sagittae is losing carbon-rich matter that condenses into carbon dust, which will hide the star until the soot dissipates. Such is the behavior of **R CORONAE BOREALIS** and others of the same rare kind. For decades, we have wondered where these weird stars came from. Now we know—from a tale told by a celestial Rosetta Stone.

A Jupiter-like planet orbits a star called HD 209458. Unlike 51 Pegasi's planet, this one crosses in front of its star, causing a light dip in starlight. *(Illustration by G. Bacon [STScI/AVL].)*

Residence:	PEGASUS
Other name:	HR 8729
Class:	G2.5 DWARF
Visual magnitude:	5.49
Distance:	59 LIGHT YEARS
Absolute visual magnitude:	4.56
Significance:	THE FIRST STAR KNOWN TO HAVE A PLANET.

51 PEGASI

We look for ourselves in the stars. Speculation about life elsewhere in the Universe is ancient. Astronomers now use giant radio antennas, among other things, to look for signs of life, for some intelligent signal from afar. But to have life, there must be planets to support it. The search for planets, while daunting, has been successful, and now dozens are turning up. With all our technology, however, we have yet to *see* one: the detections are all indirect. Nevertheless, they are convincing enough to show that planets orbit stars not too different from the Sun.

Two bodies in gravitational embrace orbit each other. The Moon orbits the Earth, but the Earth also orbits the Moon. Since the Moon has 1/81 the mass of Earth, the Earth's orbit has 1/81 the diameter of the lunar orbit and swings around a point 4750 kilometers from the terrestrial center (inside the Earth). As Jupiter, 1/1000 the solar mass, revolves around the Sun, the Sun is deflected along a path 1/1000 the size of the Jovian orbit, about 1/100 of an AU.

If a body moves toward you, its spectrum of colors is shifted toward shorter wavelengths, and if it moves away from you, the wavelengths become longer (the Doppler effect). The absorption spectra of orbiting double stars shift one way then another (as we see for **MIZAR** and other stars). If a planet orbits a star, the star's spectrum gives it away, as the absorptions will alternately move back and forth. Because the planet is so light compared with the star, the shift is minute, but technology is almost good enough to detect our Jupiter around our Sun. If the planet is closer to its star or more massive than Jupiter, it will be found.

A fifth magnitude star at the edge of the Great Square of Pegasus was first. Similar to our Sun, 51 Pegasi is a G2.5 dwarf, with a temperature of 5800 K. Only 50 light years away, it shines with 1.3 times the solar luminosity and has a mass only 5 percent greater than solar. It does have one peculiarity shared by many of the stars that have planetary companions: it is rich in metals. With 60 percent more iron than our Sun, it clearly formed from an interstellar cloud that was enriched by the actions of previous stellar generations.

Astronomers once thought that our planetary system would be exemplary, and yet, the planet orbiting 51 Pegasi is outstandingly strange. It is a Jupiter-like body with a mass at least 60 percent of our Jupiter. (Since we do not know the orbital tilt, we cannot know the actual mass, only a lower limit.) But our Jupiter is 5 AU away from our Sun and has an orbital period of 12 years; 51 Pegasi's planet is tucked in only 7.5 million kilometers away from its star and makes the orbital circuit in 4.2 days—just 5 percent that of Mercury's period. This "system" (if indeed there is more than one planet) is nothing like our own.

Planets are believed to accumulate from dusty disks that orbit new stars. "Jupiters," which consist mostly of hydrogen, cannot form that close to their parent stars because it is too hot to allow the volatile hydrogen to condense. The only explanation we have is that 51 Pegasi's planet formed far from its star and then migrated inward thanks to the frictional effects of a residual disk, something our Sun did not have. The existence of such odd planets may be related to the high metal-contents of their stars. We do not know. In such a new science, almost everything is a surprise. Does an "Earth" orbit 51 Pegasi as well? With that big "Jupiter" moving inward, it seems unlikely, but there may well be other "Earths" out there just waiting for us to find them. Maybe somebody is looking back at us.

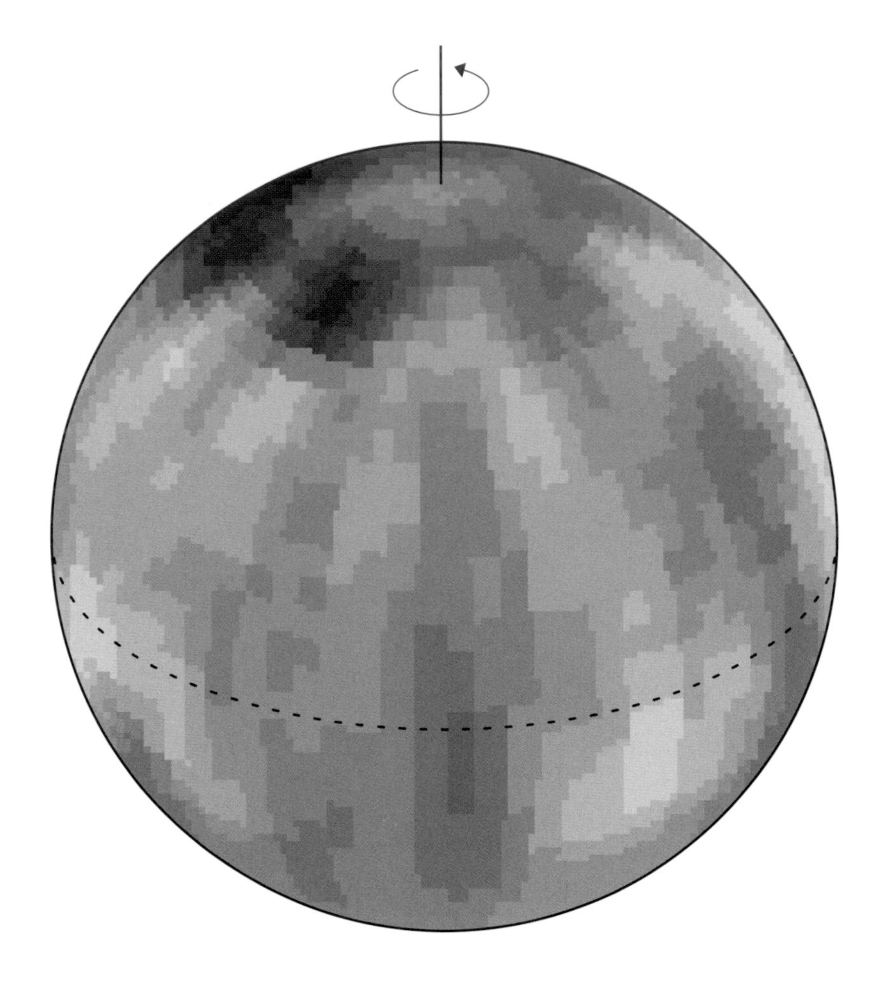

A temperature map of FK Comae Berenices, derived from Doppler (velocity) variations in the star's spectrum, reveals the surface of the star, allowing us to watch starspots go in and out of view. *(H. Korhonen et al., Nordic Optical Telescope, from an article in Astronomy and Astrophysics 346 [1999]: 101.)*

Residence:	COMA BERENICES
Other name:	HD 117555
Class:	G5 BRIGHT GIANT
Visual magnitude:	8.17
Distance:	760 LIGHT YEARS
Absolute visual magnitude:	1.32
Significance:	WILDLY SPOTTED STAR.

FK COMAE BERENICES

The most obvious feature of the Sun is its spots, which are produced by loops of intense magnetism generated by solar rotation and convection. The magnetism is also responsible for a vast, hot surrounding solar corona whose temperature reaches over 2 million K, enough to produce X-rays. Collapsing chaotic magnetic fields create bright solar flares and eject bubbles of coronal gas that produce the Earth's aurorae.

Two-thirds of the class G dwarfs exhibit similar magnetic activity. While we cannot see the spots directly, the signature of varying activity shows up in the spectral features (the absorptions and emissions produced by specific atoms and ions), as well as in variations in the stars' brightness. Some stars are more active than the Sun, others less. For extreme activity, however, it is hard to top the FK Comae Berenices stars—all five of them.

FK Comae is a G5 to G8 giant of 8 solar diameters 760 light years away. With a temperature of 5100 K, it radiates at 30 solar luminosities. What makes it so special is its immense rotation speed of 180 kilometers per second, enough to make the star begin to flatten itself out. If the Sun, rotating at only 2 kilometers per second, can produce spots and other activity, why not FK Comae? Sure enough, it is variable, changing its visual brightness by about one tenth of a magnitude over a period of 2.40 days. The variations are interpreted as collections of cool large spots that rotate in and out of view. Moreover, strong emissions from hydrogen and other atoms reveal a surrounding active "chromosphere" (which on the Sun separates the corona from the bright surface), and powerful X-rays suggest a similarly active corona.

"Doppler imaging," a process that uses changes in the line-of-sight velocities of spectral emissions and absorptions to make maps of stars, reveals spots near the poles and equator. The spotted regions can last for years, their number seeming to come and go on a ten-year cycle. The spots also display a mystifying "flip-flop" effect, in which the areas switch longitudes by 180 degrees. Activity, visible in X-rays and in the hydrogen emissions, can be extreme. One flare observed in 1997 set a record for cool stars when its hydrogen emission alone released more much energy than does the Sun in nearly one hour.

The explanation for such outstandingly odd stars sounds odder yet. Normally, stars do not collide—they are just too far apart. However, there is no reason why the members of close double stars cannot merge to become one. FK Comae Berenices strongly reminds us of **RS CANUM VENATICORUM** and its ilk. They are mid-temperature, short-period binaries in which the pairs are tidally locked to each other (that is, perpetually facing each other). As a result, they rotate quickly, generating great starspot activity. FK Comae, however, is distinctly single. It may be the merged product of what was once an RS Canum Venaticorum or **W URSAE MAJORIS** double. The two stars coalesced when one of them expanded during its early evolution and incorporated the other, the orbital energy going into single-star rotational energy. FK Comae and its rare breed thus vividly demonstrate the dynamic and intimate nature of stellar life and also help reveal the origins of the solar activity that so affects the Earth.

40 Eridani-B, to the west of Orion and tucked next to the modestly bright K1 dwarf 40 Eridani-A (and itself companion to a fainter M dwarf) is among the most famed of white dwarfs. *(J. B. Kaler; inset* © National Geographic—*Palomar Observatory Sky Survey, reproduced by permission of the California Institute of Technology.)*

Residence:	ERIDANUS
Other name:	KEID = OMICRON-2 ERIDANI
Class:	K1 DWARF + DA WHITE DWARF + M4 V DWARF
Visual magnitude:	4.43
Distance:	16.5 LIGHT YEARS
Absolute visual magnitude:	5.92
Significance:	TRIPLE, WITH THE FIRST WHITE DWARF.

40 ERIDANI

Beid and Keid, unrelated stars meaning "the eggs" and "egg shells" in Arabic, carry the same Greek letter name, Omicron Eridani. We discriminate by calling Beid "Omicron-1," Keid "Omicron-2," though astronomers know them better as John Flamsteed's "38" and "40 Eridani." Forget 38. Our target is 40 Eridani. Only 16.5 light years away and 50th-closest to the Earth, this unassuming star plays a crucial role in our understanding of stellar science.

Fourth magnitude to the eye, 40 Eridani is famed not for the naked-eye star (called 40 Eridani-A), but for a faint companion to which it is a valuable guide. Other than being a signpost, 40 Eridani-A (a K dwarf) has little to offer but a cool temperature of 5100 K, a low luminosity of 0.4 times that of the Sun, and a mass around three-fourths solar. A little over one minute of arc away lies the system's prize, 40 Eridani-B, a tenth magnitude white dwarf. Though **SIRIUS B** is better known, closer to us, and brighter, 40 Eridani-B is by far more visible, as 40 Eridani-A is so much fainter than brilliant Sirius A. Therein lies the story.

Measurement of a star's mass requires that it be part of a binary system in which gravitational theory can be applied to the mutual orbit. 40 Eri-B is sadly too far from 40 Eri-A for any orbital motion to be detected. With a minimum separation of 420 AU, the orbital period is at least 7300 years. (The star has yet to make a full orbit since Eridanus was invented!) However, the white dwarf has its own companion (40 Eridani-C), a dimmer 11th magnitude class M ordinary dwarf only 34 AU away. Though they take 248 years to circle each other, the partial orbit we have observed thus far is enough to allow a mass determination.

By the early twentieth century, Sirius B, Procyon B (the companion to Procyon, Alpha Canis Minoris), and 40 Eridani-B were all known for being faint stars with anomalously high masses—but for little else. Of these, only 40 Eri-B was far enough from its companion to be examined with the spectroscope. To everyone's great surprise, 40 Eri-B had powerful hydrogen absorptions, implying that it was a hot, white class A star. (We now know that ordinary classes do not apply to white dwarfs. Sirius B and 40 Eridani-B are both DA white dwarfs with hydrogen-rich atmospheres. The heavier helium has sunk out of sight.) To be both hot and dim, it had to be small, roughly the size of Earth. Only later, in 1915, did we find that Sirius B was similar. 40 Eridani-B thus has the honor of being the first white dwarf known, a dead star that is held up by the pressure of its fast-moving electrons.

For years, 40 Eri-B confounded astronomers. Its mass appeared too low to comply with theory. Observations by spacecraft finally allowed the calculation of an improved orbit that led to an exact mass of 0.50 solar, about half that of Sirius B. The high temperature of 16,000 K and a luminosity 1.2 percent solar infer a radius 0.0136 that of the Sun (1.5 that of Earth). These measurements match theoretical projections, thus providing a profound benchmark for evolutionary and white-dwarf theory. (Though 40 Eri-B has a lower mass than Sirius B, it has a greater radius because of lower gravitational compression.)

While now lighter than 40 Eridani-A, the white dwarf must originally have been the most massive of the three (with a mass about that of the Sun) to have evolved first. At one time, 40 Eridani-B dominated the system, especially when it was making the transition to white-dwarfhood as a giant. The K star will be the next to go, this delightful triple nicely summarizing the evolution of lower-mass stars.

FU Orionis, hidden by a disk within the imaging instrument, is surrounded by a faint dusty nebula that reflects its light. *(T. Nakajima and D. A. Golimowski, The Johns Hopkins University Adaptive Optics Coronograph, and Palomar Observatory, California Institute of Technology.)*

Residence:	ORION
Other name:	NOVA ORIONIS 1939
Class:	G0 SUPERGIANT AND PROTOTYPE VARIABLE
Visual magnitude:	9.5 (VARIABLE)
Distance:	1600 LIGHT YEARS
Absolute visual magnitude:	–1.4
Significance:	A YOUTHFUL, VIOLENTLY ACCRETING STAR.

FU ORIONIS

Stars that do something are fun to watch. In 1939, a "new star" erupted in northern Orion out of nowhere. It reached a "blue" photographic brightness approaching magnitude 9, making it easily visible in a small backyard telescope. Searches of old photographs show the perpetrator to be a faint star of 16th magnitude. Within one year, it had brightened by a factor of around 500. Novae (like the famed example Nova Herculis 1934, now known as **DQ HERCULIS**) are hardly unusual; many are found each year. They are surface explosions that take place on white dwarfs when they accumulate too much fresh hydrogen from a companion. When the fresh fuel is burned, novae quickly fade. This erupter challenged expectations by staying lit up for over 60 years. Though very slowly dimming, it still has a strong presence with a blue magnitude around 11 and a visual magnitude of about 9.5. Now known by its variable star name, FU Orionis was surely not a nova. Since then, another ten similarly behaving stars have been discovered. FU Orionis, the first, is the privileged prototype. The whole gang are now called "FU Orionis stars," or, since acronyms and abbreviations are unfortunately rampant upon the land, FUOrs. If not novae, what then?

Lurking within and around the dark dust clouds of the Milky Way are myriads of young stars, **T TAURI** stars that, within the last million years or so, have collapsed out of their local interstellar clouds and are just now seeing the "light of day." Many such stars are surrounded by dusty disks from which the new stars accrete matter. At the same time, dust particles in the disks may be accumulating into planets. Some T Tauri stars even have small illuminated nebulae around them whose dust particles reflect the stars' light. FU Orionis stars are not found in isolation, as are novae. Instead, they are located within the same kind of dusty clouds as the T Tauri stars. Moreover, they too are usually surrounded by reflection nebulae. T Tauri stars vary erratically; could the FUOrs just be an extreme of their kind?

Theory to the rescue. As a new star contracts out of interstellar matter, it spins into a disk with a central concentration. In fact, without some means of slowing the rotation, the new star would spin so fast that it could not form in the first place. The search for "star formation" is largely the search for the means to brake the rotation (the details told by **T TAURI** itself). Even when slowed enough, a disk cannot help but form. It is crucial to star formation. The growing star does not accumulate its final mass directly from space. The means for the process is instead provided by the disk, which accretes matter from the surrounding cloud at its outer edge and then dumps it to the star through its inner edge, which is effectively in contact with the stellar surface.

Usually calm and neutral, the disk is actually highly unstable. If the inflow rate is high enough, the inner edge suddenly ionizes. Massively brightening, the disk sends a huge load of matter raining down on the star. Accretion rates can increase tens of thousands of times to a ten-thousandth of a solar mass per year. Given that FU Orionis stars can be in this state for 100 years or more (none have ever been seen to return to normal), and that they may undergo such events periodically, a large fraction of the final stellar mass may be accumulated in this way. The "nova" effect is not so much from the star, but from the disk, the star acting as the constraining force that keeps the disk together. The Sun probably went through similar events as it was developing. As with people, outrageous and unpredictable adolescence seems necessary for maturity.

Gamma Cassiopeiae (see also #95) shines brightly at the center of Cassiopeia's "W," seen here below the pole above the Black Hills of South Dakota. BETA CASSIOPEIAE is up and to the right of Gamma, while RHO CASSIOPEIAE (see also #95) is just lost in the trees to the right of Beta. *(Rick Olson.)*

Residence:	CASSIOPEIA
Other name:	27 CASSIOPEIAE
Class:	B0 SUBGIANT WITH EMISSION
Visual magnitude:	2.47 (1.6 TO 3.0)
Distance:	615 LIGHT YEARS
Absolute visual magnitude:	–4.4
Significance:	THE FIRST KNOWN "B-EMISSION" STAR

GAMMA CASSIOPEIAE

Like opposing hands of a clock, the Big Dipper of Ursa Major and the "W" of Cassiopeia's Chair wheel about the North Pole. The Dipper's stars are mostly named after their placement in the Greater Bear, but there are fascinating exceptions that tell of other lore. So it is with Cassiopeia, the Queen, the Lady in the Chair. Shedar, her Alpha star, means "the breast," while Ruchbah, Delta Cassiopeiae, refers to her knee. Beta, however, is Caph, "the stained hand" (referring to a traditional Arabic practice of cosmetic dying with henna), the hand composed of "fingertips" made of Alpha through Epsilon Cassiopeiae. Caph then went to Delta alone, leaving poor Gamma with no name at all, the star a victim of early collectivization. It is arguably the most prominent star in the sky, right at the center of the famed "W," with no common name. And what a pity, as it is not only a magnificent star but also the exemplar of its kind. In the 1860s, Father Angelo Secchi developed a system that described four types of stellar spectra. The first included stars like **SIRIUS** that have powerful hydrogen absorptions, the second those similar to the Sun, and then two kinds with different molecules that respectively include reddish stars like **BETELGEUSE** and deep red carbon stars like **R LEPORIS**. The system led directly to Harvard's famed OBAFGKM sequence of 1891. But Secchi saw few stars like Gamma Cas, whose spectrum was so outlandish that he created a whole new "fifth type" for it and **BETA LYRAE**. Gamma Cas's spectrum has the absorptions we now relate to hot class B, but also has strong hydrogen emissions ("e"), making it the first known "Be star." We now know such stars to be common, the class including one of the horns of Taurus (Zeta Tauri) and Alcyone, the luminary of the Pleiades, as well as others of the "Seven Sisters."

Beta Lyrae's oddness comes from its duplicity, while Gamma Cas's (and that of the other Be stars) comes from spin of immense proportions. All stars—indeed all celestial bodies—rotate. But how can we find the spin of a star that appears as a point? In some cases, spots on the stars roam in and out of view, making the star vary in brightness. In others, spectra give it away. As a star rotates, one side comes at us (never mind the star's gross motion), its absorptions Doppler-shifted to shorter wavelengths, while the other side moves away, red-shifting the lines. Averaged over the star, the absorptions of a fast spinner are broad, while those of a slow one are narrow. Cool main sequence stars spin like the Sun. But as we go to higher masses, luminosities, and temperatures, around class F the rotation speeds climb rapidly, peaking in class B. Gamma Cas rotates at an astonishing 300 kilometers per second, and maybe more, as we do not know how the axis might be tilted. The forces at the stellar equators are so great that Be stars must severely flatten out and come close to the point of breaking up. When coupled with winds from high luminosities, Be stars become surrounded by thick, spinning disks of illuminated gas that produce the emissions.

And Gamma Cassiopeiae is different yet. It is an unpredictable variable that reached almost first magnitude in 1937 and then faded to third, the variations clearly related to its unstable rotation-driven surroundings. Perhaps its lack of a proper name tells us of ancient dimness. The star also radiates X-rays, though no one is quite sure why. Theories include the transfer of lost mass to a compact companion and magnetic effects similar to those found on the Sun. If any star deserves a name, surely this one does.

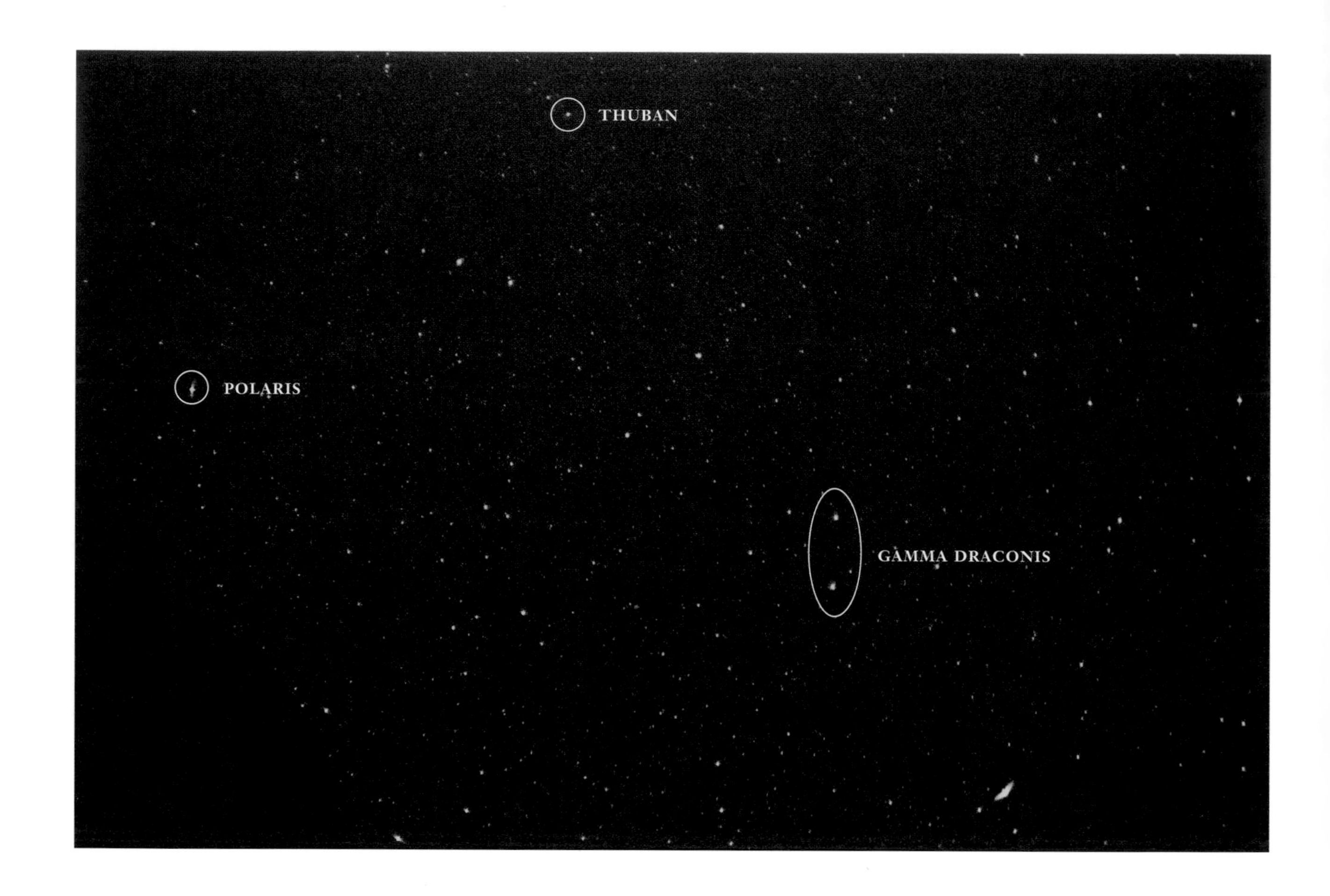

The two eyes of Draco (to the lower right of center), Gamma Draconis, the lower of them, looks to the right toward Hercules. THUBAN, the pole star of ancient Egypt, is to the left of top center, just up and to the right of the bowl of the Little Dipper. *(J. B. Kaler.)*

Residence:	DRACO
Other name:	ELTANIN
Class:	K5 GIANT
Visual magnitude:	2.23
Distance:	148 LIGHT YEARS
Absolute visual magnitude:	–1.05
Significance:	THE HISTORIC “ZENITH STAR.”

GAMMA DRACONIS

Since the Copernican Revolution in 1546, we have known, or at least surmised, that the Earth is not stationary and at the center of the Universe, but instead spins on its axis and orbits the Sun. Proving these "truths" is another matter. Gamma Draconis helped pave the way.

At a distance of 148 light years, this second magnitude orange giant radiates 590 solar luminosities into space from a relatively cool 4000-K surface. A tenth of a magnitude variation may be the precursor to the star's eventual conversion to a pulsating **MIRA**-type variable. The star's claim to greatness lies not in its substance, however, but in its position. Eltanin is the closest bright star to the "winter side" of the "solstitial colure" (the circle in the sky that passes through the celestial poles and the solstices), lying almost 75 degrees north of the winter solstice in Sagittarius. Eltanin's high northerly position of 51.5 degrees north of the celestial equator takes it through the zenith, the point overhead, as seen from London.

In the 1700s, the London area was a center for the measure of precise stellar positions. One of the major difficulties in establishing them is the refraction of light caused by the Earth's atmosphere, which makes stars appear higher in the sky than they actually are. Even after correcting for refraction, astronomers found that they still could not get consistent positional measures. The key was to eliminate refraction by observing a star that passed directly overhead: Gamma Draconis!

The star seemed ideal to use in the search for parallax, and therefore distance. And indeed, Gamma Draconis refused to "stand still," shifting back and forth seasonally, but in the wrong direction for parallax to be the culprit (the discovery of which had to wait until 1838 for Bessel and **61 CYGNI**). Instead, in 1728, James Bradley announced that the star was shifting as a result of the movement of the Earth and the finite velocity of light. As you walk into a snowstorm on a windless day, the flakes still seem to come from a point in front of you. Similarly, as the Earth moves perpendicular to "falling" starlight, the star seems to be shifted in the forward direction. The degree of shift depends on the speed of the Earth (rather the component of the speed perpendicular to the direction to the star) relative to the velocity of light and has a maximum of 20.5 seconds of arc, theory and observation corresponding exactly. This "aberration of starlight" was the first real proof that the Earth truly moves and orbits the Sun.

Bradley was not yet done with the star. The Moon and the Sun act on the Earth's equatorial bulge to produce a 26,000-year axial wobble, or "precession," that changes the pole star and the visibility of the constellations. Further observations of Gamma Dra led to the discovery of "nutation," a 17-second-of-arc sub-wobble superimposed on the precession caused by variations in the lunar orbit. Over 100 more nutational terms were later found. When all these effects are accounted for, we can precisely determine stellar positions.

As with people, it sometimes does not matter *who* you are, but rather *where* you are.

Wolf-Rayet stars are known for their ejecta, which are exemplified by the complex shell known as NGC 6888, which surrounds a WN star. Perhaps Gamma-2 Velorum will someday pop such an envelope. *(Hubble Space Telescope, B. D. Moore, J. Hester, P. Scowen [Arizona State University.], R. Dufour [Rice University], STScI, and NASA.)*

Residence:	VELA
Other name:	REGOR
Class:	WC8 WOLF-RAYET + O 7.5 GIANT
Visual magnitude:	1.78
Distance:	840 LIGHT YEARS
Absolute visual magnitude:	–5.28
Significance:	A DOUBLE WITH THE BRIGHTEST WOLF-RAYET STAR.

GAMMA-2 VELORUM

Remember the **ALGOL** paradox, in which the more-evolved star that should be the more massive, is actually the less massive? Welcome to the extreme. Gamma-2 Velorum (a much fainter class B companion to the west receiving Gamma-1) is a bright, blue-white second magnitude star within Vela, the sails of the great ship Argo. (The star is Vela's luminary and the vessel's number 3, the Greek letters distributed across Argo's three parts.) The Arabs referred to several of Argo's stars that skimmed their southern horizon as "Suhail," including **CANOPUS** and Gamma-2. Since Lambda Velorum is more commonly called Suhail, Gamma-2 is demoted to the more modern Regor, "Roger" spelled backwards. No one knows what *Suhail* actually means. "Regor," however, honors Roger Chaffee, who was one of the three tragic fatalities in the Apollo lunar-landing program.

The star—"the spectral gem of the southern skies"—is spectacular, its visual luminosity alone 11,000 times solar. Of much more interest, however, is that the star is double. Taking only 76 days to orbit each other, the components are inseparable except by means of sophisticated interferometry. However, the duplicity is also evidenced by a composite spectrum, that of an ordinary class O giant (as if such are ordinary!), combined with the bizarre spectrum of a Wolf-Rayet star.

Co-discovered by French astronomers Charles Wolf and Georges Rayet in the nineteenth century, "WR" stars have none of the usual absorption lines, but instead are filled with bright spectral emissions. There are two kinds: "WC" stars that are loaded with emissions from helium and carbon, and "WN" stars that shine the spectra of helium and nitrogen. The twain do not meet or merge; WR stars are one or the other, never both. Moreover, WC stars have no hydrogen, the most common stellar element. WR stars are rare, hot, very blue, wonderfully luminous, and madly losing mass. Regor's WR member is a WC star and, visually, the brightest of any of them. Relative to helium, a WC star's carbon ratio is 100 times more than normal. There is little question about the reason for such an odd chemical composition. The WC star must have lost its entire outer hydrogen envelope to reveal the by-products of nuclear fusion that were created in the deep core. In the WN stars, we see the effects of hydrogen fusion that has altered the nitrogen and helium mixture. Here, we see the results of a more advanced stage in which helium was fused into carbon.

Regor's impressive brilliance in the visual spectrum is but a beginning. Both members radiate most of their light in the invisible ultraviolet. The O giant is the brighter, pumping 200,000 solar luminosities from its 34,000-K surface, to which between 50,000 and 100,000 more "Suns" are added from its much hotter (an uncertain 60,000 K) WC neighbor. Analysis of the orbit combined with theoretical evolutionary calculations show that the O star contains around 30 solar masses, while the WC star contains only between 5 and 10. To have evolved first, the latter must have been bigger than the O star. The WC star has lost 20 to 30 times the mass of the Sun into space, which accords well with the weird chemical composition and the fierce wind from the stellar surface. Even now, the wind blows at a rate of nearly a ten-thousandth of a solar mass per year, 1 billion times the flow in the solar wind. The wind carries with it the carbon that was created through helium fusion. We can watch the enrichment of the interstellar gases, the material from which new stars are made, happen before our eyes. Maybe some of our own carbon came from just such a star aeons ago.

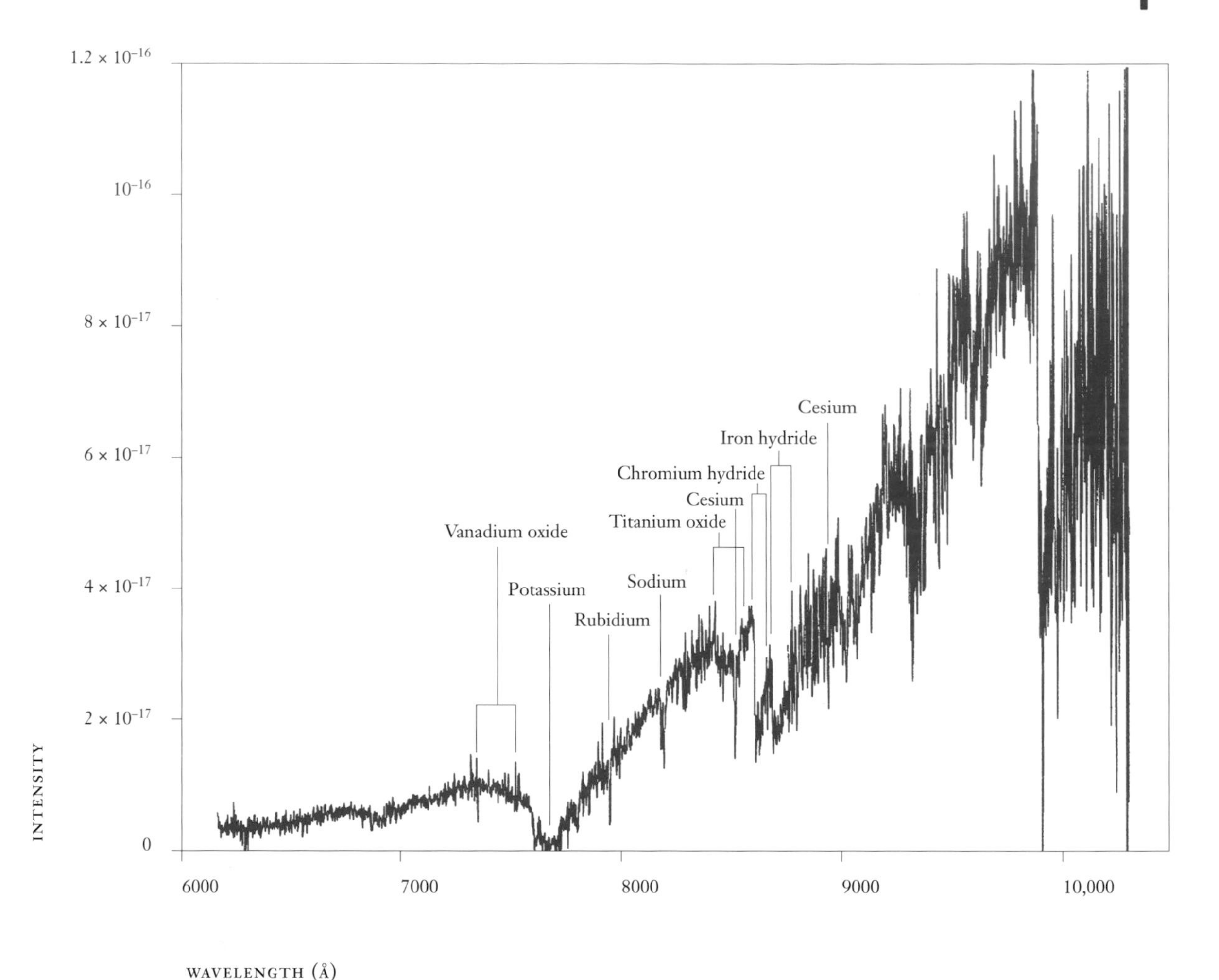

The red and infrared spectrum of the cool brown dwarf GD 165B reveals broad absorptions caused by metal oxides and hydrides, as well as powerful narrow dips created by the presence of the alkali metals sodium, potassium, rubidium, and cesium. *(W. M. Keck 10-meter [California Institute of Technology, University of California, and NASA] spectrogram, courtesy of J. D. Kirkpatrick et al., from an article in the* Astrophysical Journal *519 [1999]: 834.)*

Residence:	BOÖTES
Class:	L4 BROWN DWARF
Visual magnitude:	24
Distance:	103 LIGHT YEARS
Absolute visual magnitude:	21.5
Significance:	THE FIRST SUSPECTED BROWN DWARF AND A PROTOTYPE FOR CLASS L.

GD 165B

As main sequence masses decline from one solar, luminosities drop precipitously. Cooler than about class K7 at a mass just over half solar, stars become so dim that they disappear from the naked-eye sky. As the mass drops, the gravitational compression and the interior temperature also decrease, leaving low-mass stars unable to generate much energy.

Main sequence stars are supported and kept from using their gravitational energy by fusion, as ordinary hydrogen is converted into helium. The process requires that hydrogen atoms smash together at high velocities. At a mass of 0.075 solar, the temperature (around 7 million K) is not sufficient to generate the required atomic speeds, and the star cannot run the full fusion chain. We have reached the bottom of the main sequence, at spectral class M9 or M10, where stars are visually 1 million times fainter than the Sun. (They are more respectable when we add in the invisible infrared radiation.)

The search for stars below this limit, for "brown dwarfs," was long and arduous and included many candidates, most of which were dropped. Among the best was one that finally made the grade, GD 165B. GD 165B was discovered hiding 4 seconds of arc away from the relative glare of a 17th magnitude white dwarf named Giclas Degenerate (GD) 165 ("degenerate star" is another name for "white dwarf"). At its distance of 103 light years, the pair must be more than 125 AU apart and take over 1700 years to orbit.The spectrum of GD 165B was found to be different from any yet observed. While low mass class M stars are defined by absorptions of titanium and vanadium oxide molecules, GD 165B has almost no vanadium oxide and only weak titanium. Instead, it is loaded with hydrides of iron and chromium, a violation of everything we thought we knew about the spectra of stars. To have such a spectrum, the temperature must be only about 1900 K, the cool surface and small size rendering the luminosity a mere 5 millionths solar.

The "brilliant" white dwarf neighbor helps explain this oddity. The spectrum of this hot (12,500 K) variable "**ZZ CETI**" star reveals a mass of between 0.56 and 0.65 solar which, in turn, suggests an initial mass between 1.2 and 3 solar masses (the "missing mass" lost through its giant-stage wind). It takes between 1.2 and 5.5 billion years for such a star to turn into a white dwarf. As low-mass main sequence stars develop, they dim and, upon firing their internal hydrogen, stabilize. Brown dwarfs below the limit just get continuously dimmer. From the white dwarf's age and the brown dwarf's luminosity, the brown dwarf's mass is right below the edge, 0.073 solar. If GD 165B is not a genuine brown dwarf, it is at least a "transition object" that might eventually fire its hydrogen. Given the increasing slowness of evolution with dropping mass, however, this would not happen until after a time greater than the age of the Universe. Either way, the star is a centerpiece of brown-dwarf research. Complicating the issue (and there always seems to be something complicating the issues), the more massive white-dwarf companion might actually be a double white dwarf in the mold of **LB 11146**, which compromises some of the conclusions.

New infrared detectors are uncovering vast numbers of such stars, so many that a whole new spectral class "L," which is cooler than class M, had to be created. The hotter end of the sequence is probably a mixture of real stars and brown dwarfs, whereas the cooler end is certainly all brown dwarfs. GD 165B comes in at L4, close to the middle. The little star acts not only as a superb flagship for its class, but also provides a fine entry to an even dimmer stellar netherworld epitomized by **GLIESE 229B.**

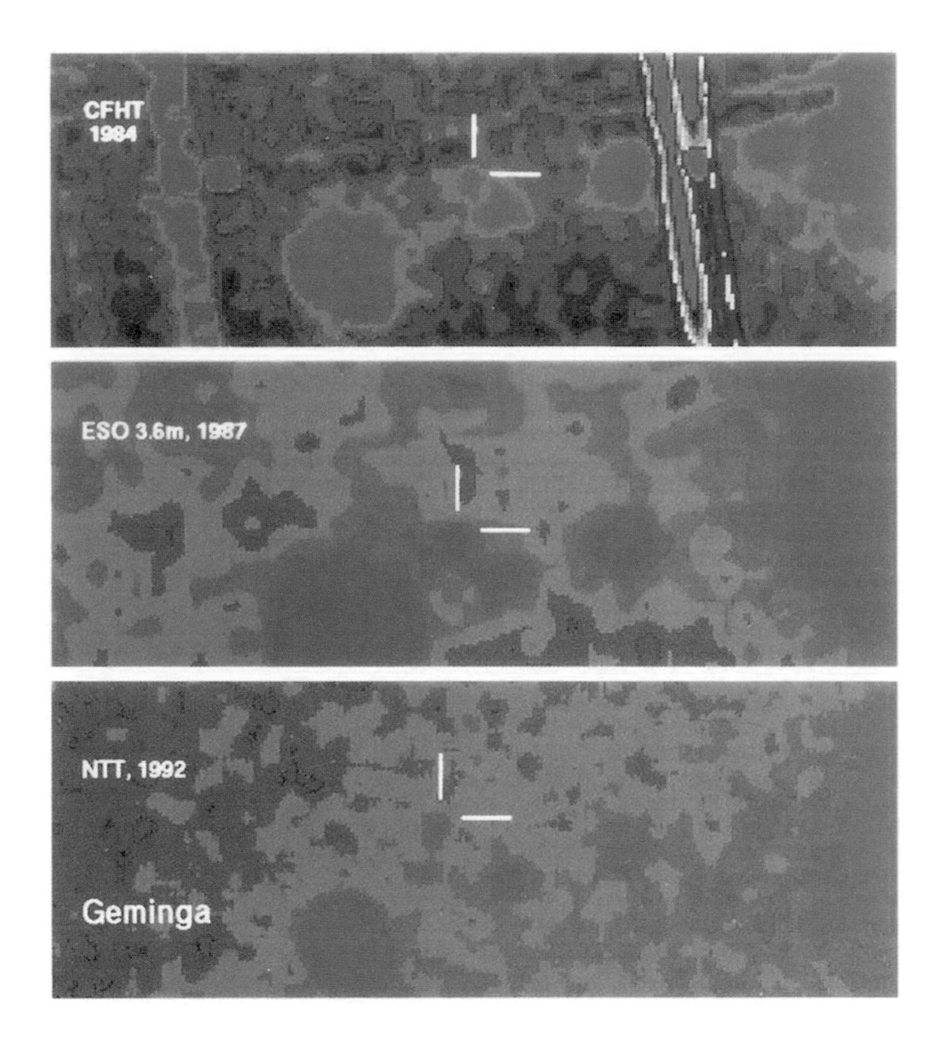

Three pictures, taken with different telescopes, show the motion of the nearby but exceedingly faint Geminga pulsar. *(Top: Canadian-France-Hawaii Telescope. Bottom two: European Southern Observatory. From an article by G. F. Bignami and P. A. Caraveo that appeared in the* Annual Review of Astronomy and Astrophysics *34 [1996]: 331.)*

Residence:	GEMINI
Other name:	G″
Class:	NEUTRON STAR/PULSAR
Visual magnitude:	25.5
Distance:	800 LIGHT YEARS
Absolute visual magnitude:	18.5
Significance:	A RADIO-QUIET GAMMA -RAY PULSAR.

GEMINGA

Think "pulsar" and think "radio." These rotating neutron stars were first found in that low-energy radiation band. A few, however, led by the young **CRAB PULSAR**, spin so quickly and energetically that they radiate not just in the radio but also in the optical, the X-ray, and even in the immense-energy gamma-ray domains. These dense neutron stars, only 20 or so kilometers across, do not actually "pulse." Instead, they radiate energy along magnetic axes that are tilted against the rotational axes. As a pulsar spins, the magnetic axis wildly wobbles, and if Earth is in the way of the passing beam, we get a shot of energy. The magnetic fields are huge, 100 million times that of Earth. The pulses essentially come from the magnetic fields through the energy of rotation. As pulsars age, they lose both their energy and their pulses from the top down. The high-energy gamma rays disappear first, followed by the lower energy X-ray and optical, until nothing is left but radio.

It all made perfect sense until one object turned the whole field upside down. In 1973, an orbiting satellite found a bright gamma-ray source in the constellation Gemini that had no obvious counterpart. Further observations showed that the object was pulsing at X-ray energies with a typical young-pulsar period of 0.237 seconds (4.2 times per second). Perversely, the object was emitting neither radio nor optical waves. Because the curious critter was in Gemini, it acquired the nickname "Geminga," which, in the Milanese dialect of Italy means "not there," a marvelous reference to its seeming optical invisibility. Deep imaging showed a few stars in the field, the most obvious labeled "G" for "Geminga." As the location of the high-energy pulses got better, the search was narrowed first to a fainter star called G′ and then to an even fainter one near the limit of visibility called G″. All these efforts were complicated by the motions of the stars relative to one another and by the temporary invasion of a faint rogue asteroid. Timing measurements of the gamma rays against the motion of G″ finally nailed it down—a dim, dim star of magnitude 25. At high energies, Geminga behaves like a real pulsar. At optical wavelengths, it appears as a bare neutron star, one of a very few known, radiating its own heat from a surface with a temperature between 200,000 and 300,000 K.

Neutron stars are the progeny of supernova explosions, the ultradense collapsed cores of massive stars. The pulse, or rotational period and the rate at which Geminga is slowing down yield an age of around 340,000 years. Where might the star have originated? Tracking its motion backward points to the complex region around Lambda Orionis, which marks Orion's head. This massive double star is in the middle of a small cluster and is centered on a mighty ring of gas 150 light years across that is set within an even bigger ring of molecular gas and dust, suggesting that a star exploded there. Such stellar explosions, for unknown reasons, seem to take place off-center, causing the resulting collapsed neutron stars to go reeling outward at high speed. Geminga, moving across the line of sight at 120 kilometers per second, is perhaps the result.

But what of the radio waves? In spite of intense searches, there do not seem to be any. Perhaps the high-energy radiation has a broader "beam" than the radio so that it swipes the Earth with each rotation, whereas the radio does not. Perhaps there *isn't* any radio radiation. Nobody knows. In the radio, Geminga is still *geminga*.

Gliese 229, on its own a dim red class M1 dwarf, is overwhelmingly brilliant compared with the class T "methane brown dwarf" that accompanies it. The spike on the bright star is an artifact of the telescope, as is its seemingly extended "disk;" in reality the star appears as a point. *(Hubble Space Telescope, S. Durrance and D. Golimowski [JHU], STScI, and NASA.)*

Residence:	LEPUS
Other name:	NONE
Class:	T, BROWN DWARF
Visual magnitude:	FAINTER THAN 25
Distance:	18.6 LIGHT YEARS
Absolute visual magnitude:	UNKNOWN
Significance:	FIRST KNOWN (AND ULTRACOOL) BROWN DWARF.

GLIESE 229B

Gliese 229 (catalogued by Heidelberg's Wilhelm Gliese in 1969) is an ordinary, ninth magnitude (9.36) class M1 star. Within the classic spectral sequence (OBAFGKM), over 70 percent of stars are class M. The M stars—so faint that none are visible to the naked eye—range from around 0.5 solar masses down to the lower limit of 0.08 solar. At the low end, full thermonuclear fusion ceases, defining the end of the main sequence. However, the processes guiding star formation know nothing of fusion, so why should Nature stop making bodies below this limit? For years, the search was on for stars below the cutoff, for failed stars, for substars, for "brown dwarfs." Many times were they discovered. And just as many times undiscovered.

That is, until Gliese 229 was examined both at Palomar and with the Hubble Space Telescope. Both telescopes showed it to have an extraordinarily faint red companion more than 10,000 times dimmer than Gliese 229 proper (now Gliese 229A). A true brown dwarf at last! The star seems to have as much in common with Jupiter as it does with an ordinary star. Its spectrum, which shows absorptions of methane (which is present in Jupiter's clouds), reveal a temperature of only 950 K, comparable with that of a self-cleaning oven. You notice that an oven on the self-cleaning cycle does not glow in the dark. Neither does Gliese 229B. Its "red" magnitude is 24.6, but at visual wavelengths, the star is quite invisible. The mass is estimated to be between 0.03 and 0.055 that of the Sun, or 30 to 60 times that of Jupiter.

The total luminosity, almost all in the infrared, is a mere 6 millionths that of the Sun. Were we orbiting such a star, we would have to be only 0.0025 AU away (half the solar radius) to get the same level of heat that we receive from our Sun! Given Gliese 229B's low luminosity and temperature, the radius must only be one tenth solar, so an orbit would be possible. But what a world it would be. At such a tiny distance, Gliese 229B would appear 20 degrees across, and the "year" would be only five of our hours long! Surely nothing like that exists (but who knows?).

At about the same time Gliese 229B was found, new infrared technologies allowed the discoveries of yet more brown dwarfs. Now, substars seem to be everywhere, so many and so strange that two new spectral classes needed to be added to encompass them. The original end of the main sequence, class M, is characterized by absorptions of titanium and vanadium oxides. Below around 2000 K (class M9 or so), these absorptions disappear and are replaced by the metallic hydrides of new class L. Though the numbers are very uncertain, there may be twice as many L stars (most with no companions) as M stars. They seem to be a mixture of real stars above the fusion mass limit and of true brown dwarfs below it. At 1000 or so K, class T, indicated by methane, takes over. Gliese 229B is a T star.

Where the brown-dwarf category stops, nobody knows. Though they cannot fuse ordinary hydrogen to helium, they are hot enough inside to fuse their natural deuterium, the heavy form of hydrogen. Even this fusion stops in bodies with masses below 13 Jupiters. Reflexive motions of stars such as **51 PEGASI** reveal the existence of Jupiter-mass planets that range upward past the 13-Jupiter limit. Planets are made from the ground up, from the accumulation of dust and gas left over after stars form. Brown dwarfs are made from the top down, directly from the gases of interstellar space. Or so we think. Can planets be made large enough to be called "stars" (albeit failed ones)? Can brown dwarfs be made small enough to be called "planets? Do they overlap? As usual, nobody knows.

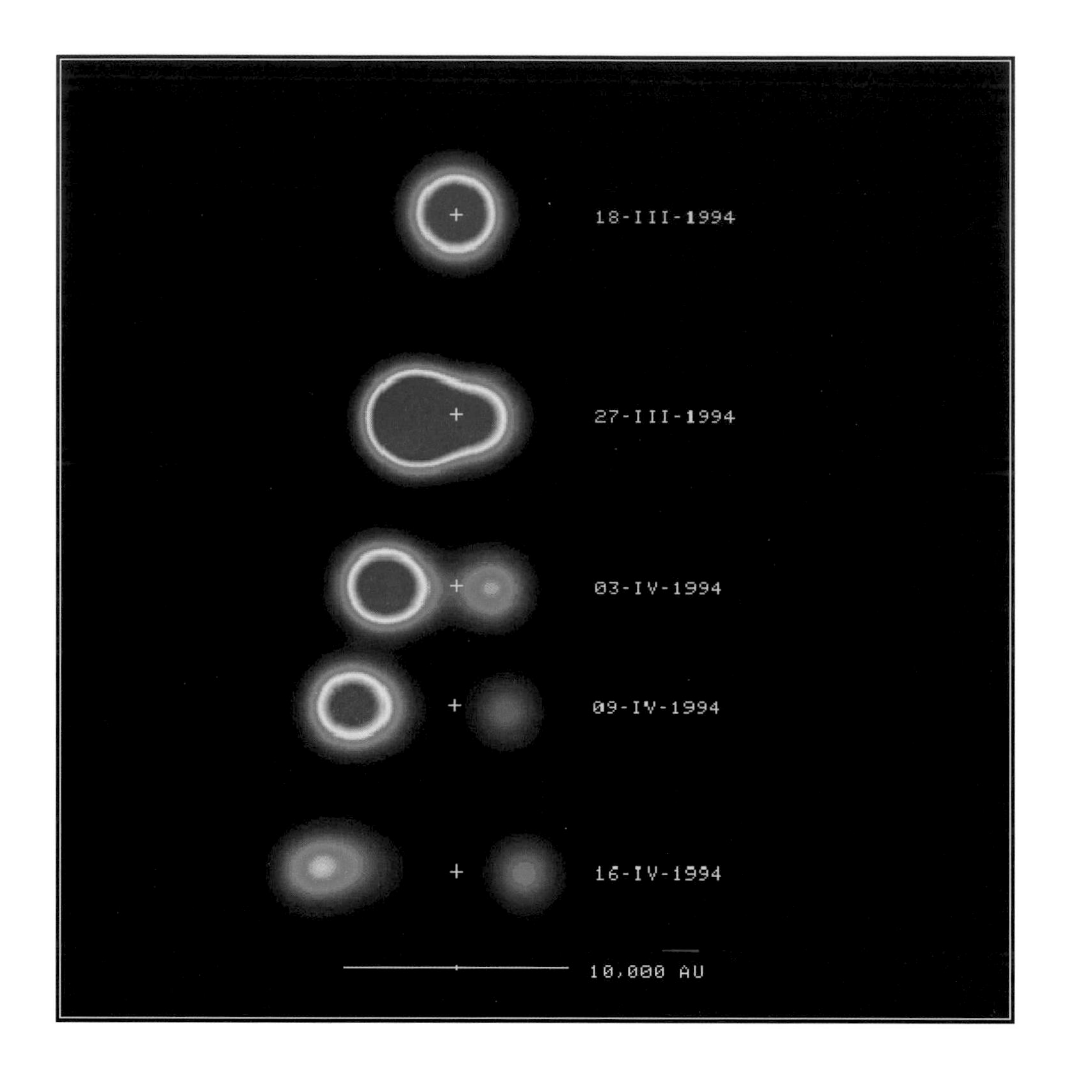

Blobs of gas move away from GRS 1915+105 at 90 percent the speed of light. An illusion caused by relativity makes them seem to move at greater than light-speed. *(Very Large Array, National Radio Astronomy Observatory, courtesy of F. Mirabel and L. Rodruguez, from an article in* Nature *371 [1994]: 46.)*

Residence:	AQUILA
Other name:	V1487 AQUILAE
Class:	BLACK HOLE MICROQUASAR
Visual magnitude:	INVISIBLE
Distance:	40,000 LIGHT YEARS
Absolute visual magnitude:	UNKNOWN
Significance:	THE FIRST "SUPERLUMINAL" GALACTIC SOURCE.

GRS 1915+105

In 1963, astronomers discovered a blue "star" whose spectrum contained emission lines that were shifted far to the red, indicating that it was rapidly receding. This star was also related to a known source of radio waves. The discovery of many like objects revealed a whole class that just looked like stars, making them "quasi stellar radio sources," or "quasars." Though most of these strange objects are radio-quiet, the name "quasar" has stuck. And though mysterious for years, we now believe that quasars are the massive black-hole cores of young galaxies that shine as matter falls into them. They are terribly far away, created in the Universe's younger days, and are taking part in the Universe's general expansion. Black holes are also thought to be the ultimate state of stellar evolution, at least of the most massive stars. If a neutron star (like the **CRAB PULSAR** or **GEMINGA**) could increase to more than about three solar masses, its neutrons could no longer support it, and it would collapse into a black hole.

Every astronomical body has an "escape velocity," the speed needed so that a launched vehicle cannot be dragged back by the force of gravity. A black hole is a collapsed—or forever-collapsing—body so dense that it is surrounded by a spherical surface (the "event horizon") at which the escape velocity equals the speed of light. As a result, the body itself disappears from view, although its gravity remains. Black holes can actually be perversely bright if they are in double systems and drawing matter from the companion. As the stuff spirals into the hole, it heats to X-ray-producing temperatures and creates bodies like **CYGNUS X-1**.

But not all the matter that falls toward a black hole needs to disappear. Supermassive black holes at the cores of galaxies are orbited by millions of stars. If any stars get too close, they can be tidally disrupted and their mass drawn inward. Intense magnetic fields then whirl some of the mass outward in powerful opposing jets that can stream for millions of light years. In a small number of cases, knots in these jets seem to be moving outward faster than light, a theoretical impossibility. Astronomers quickly realized that such prohibited "superluminal" motion is an illusion. If the jets are moving with a component that is toward us and close to, but still under, the speed of light, they will (from Einstein's theory of relativity) seem to be moving superluminally.

It would seem that huge masses would be required to perform such feats. Not so, says GRS 1915+105. Discovered by satellite from its intense X-ray emission, GRS 1915+105 is also a powerful radio source. From the interstellar hydrogen absorption line superimposed on its spectrum near a wavelength of 21 centimeters, it is estimated to be on the other side of our Galaxy, a whopping 40,000 light years away. At that distance, its X-ray emission alone would be nearly 1 million times solar. Obscuring interstellar dust produces 30 magnitudes of absorption, rendering it optically invisible. Radio and X-ray penetrate, however. Detailed radio pictures show GRS 1915+105 to be ejecting matter at apparent superluminal speeds. It is a "microquasar" in action.

A supernova in a double system probably produced a stellar-mass black hole that is now in orbit around an ordinary main sequence star from which the black hole is drawing mass. Matter first flows into a disk and is then accreted by the black hole. The X-ray emission plummets when the disk is suddenly emptied. When relativity is taken into account, twin beams are seen to flow away from the black hole at 92 percent the speed of light, the physics of a whole galaxy scaled down to the size of a star.

HD 93129A, the brightest star in the cluster toward the upper right hand corner, lies against the vast Carina Nebula along with famed ETA CARINAE, the bright star toward the lower left. *(© Anglo-Australian Observatory, photograph by David Malin.)*

Residence:	CARINA
Other name:	NONE
Class:	O3 BRIGHT SUPERGIANT*
Visual magnitude:	7.3
Distance:	9000 LIGHT YEARS
Absolute visual magnitude:	–6.7
Significance:	POSSIBLY THE MOST LUMINOUS STAR.

HD 93129A

Northern Hemisphere astronomers despair upon hearing of the glories of the deep south—the part of the sky visible from the tropics on down. Here we find the heart of the Milky Way, the Southern Cross, the Magellanic Clouds, and the **ETA CARINAE** complex. So overwhelming is the huge Eta Carinae nebula and its famed star that we might ignore another wonder, one not quite visible to the naked eye: HD 93129A.

Hardly modest at all, HD 93129A may be the most luminous star of the Galaxy. The setting is spectacular. The star is the luminary of a dense young cluster called Trumpler 14, which contains hosts of O stars, including three extra-massive O3s that lie at the hot end of the classification scheme. Two of these O3 stars form the binary HD 93129. The A component is our O3 Ia supergiant; HD 93129B, an O3 main sequence dwarf 1.5 magnitudes fainter, lies 3 seconds of arc away.

The distance of Trumpler 14 and its stars is estimated to be around 9000 light years. Even so, at a magnitude of 7.3, HD 93129A is almost visible to the eye. Were it not for 1.7 magnitudes of absorption caused by intervening interstellar dust, it would be. From its distance, corrected for the pernicious and uncertain effects of the dust, the absolute visual magnitude is –6.7. If it were as close as **VEGA**, it would far outshine Venus in our sky. From HD 93129B, which lies at least 8000 AU from HD 93129A, it would appear perhaps magnitude –18, over 100 times brighter than our full moon. (Though no orbital motion has been seen, the two most likely form a true double with a period of at least 50,000 years).

The visual magnitude is only part of the story. The star is so hot (52,000 K) that most of its radiation pours out in the ultraviolet. The "bolometric magnitude," which accounts for all of the radiation, is estimated to be around 4.6 magnitudes brighter, giving us a spectacular total absolute magnitude of –11.2. Comparison to the Sun's absolute bolometric magnitude of +4.75 yields a total luminosity 2.5 million times that of the Sun, somewhat less than Eta Carinae's. Given the inevitable uncertainties in distance and dust absorption, HD 93129A could in fact top Eta. To orbit such a star and keep the same radiation level as we have on Earth, we would have to be 1600 AU away from it, 40 times Pluto's distance from the Sun. But HD 93129B would probably make such a planet impossible. If B were not there, this hypothetical "Earth" would take some 4800 years to make an orbit; imagine a winter of 1200 years.

Such great luminosity can come only from a huge mass. Though it is now losing matter at a rate of 1/200,000 of a solar mass a year, its birth weight must have been in excess of 120 solar masses. Even HD 93129B, with a luminosity of 1 million or so Suns, must contain around 80 solar masses. The mass outflow from HD 93129A produces emission lines in the spectrum. Shock waves in the violent wind, which is 20,000 times more massive than the solar wind, create X-rays that tell of temperatures in the millions of K.

As the more massive of the pair that make HD 93129, HD 93129A was fated to evolve first, and as a supergiant now 20 times larger than the Sun has clearly started its death cycle. Less than 2 million years old, its only course seems to be to explode as a supernova, maybe as a so-called ultra-violent "hypernova." Too massive (or so we believe) to make a neutron star, the remains of the explosion will likely fall into a black hole of its own making. HD 93129B will shortly follow, perhaps making a double black hole.

The surface layers of some white dwarfs are nearly pure helium. Made in abundance in the Big Bang (the event that created the Universe) and manufactured in stars, helium is actually very rare on Earth. *(J. B. Kaler.)*

Residence:	COMA BERENICES
Other name:	WD 1211+332
Class:	DO WHITE DWARF
Visual magnitude:	14.22
Distance:	196 LIGHT YEARS
Absolute visual magnitude:	10.33
Significance:	HOT HELIUM WHITE DWARF.

HZ 21

The end of the line for stars under ten solar masses is the white dwarf, the stripped nuclear burning core of a once-powerful hydrogen-fusing main sequence star. When the solar core turns entirely into helium, it will shrink and heat, eventually firing up to fuse itself to carbon and oxygen. Simultaneously, the star will brighten and its outer parts expand, first creating a giant star like **ARCTURUS** and then, when the helium-burning has ceased, an even larger giant like **MIRA**. Such huge and luminous stars easily lose their outer envelopes. During the planetary-nebula stage (exemplified by **NGC 6543**), the white dwarf within is revealed, a star that has squeezed half a solar mass or more (up to a limit of 1.4 solar) of carbon and oxygen into a body the size of Earth.

White dwarfs are essentially cinders. Supported by the outward pressure of their electrons, their thermonuclear fusion shut down long ago. They cool from a maximum of about 100,000 K to about 3500 K (the coolest now known), while dimming at a constant radius that depends on mass. It all seems so simple until one looks at the spectra, where the mysteries reveal themselves. White dwarfs break into two sequences. The first white dwarfs found had hydrogen in their spectra, so they were thought to be like ordinary class A stars and acquired the name "DA white dwarfs." The other kind exhibited neutral helium and, in obeisance to the B stars, were called "DB." The spectra of some hotter stars contained ionized helium. Looking something like O stars, they were called "DO." Real O and B stars have hydrogen, lots of it, just as real A stars have abundant quantities of helium. The DA and DB + DO varieties, however, are pure. The DA have pure hydrogen outer atmospheres, the DB and DO pure (or nearly so) helium.

The majority of white dwarfs are DA stars. The atmospheres are so quiet that whatever helium (and heavier elements) exists is simply drawn downward by great gravity, leaving nothing at the top but helium. Among these, we find famed **SIRIUS B** and **40 ERIDANI-B**. HZ 21 (after Milton Humason and Fritz Zwicky), however, is a prototypical DO star and beautifully represents the other sequence of DO and DB white dwarfs, which are commonly lumped together as "non-DA." Shining at tenth magnitude from a distance of 196 light years, HZ 21 radiates 40 percent the luminosity of the Sun from a hot, 53,000-K surface. Measures of both the surface gravity and the luminosity reveal a radius only 80 percent of Earth's and a mass 0.56 times of the Sun, yielding an average density of 1.5 tons per cubic centimeter. The non-DA stars seem to have lost their hydrogen to space through winds. There simply is none, or at least very little. Mystery explained.

Not really. We find DA stars all along the cooling sequence. But not the DBs. There are none at all between temperatures of 30,000 and 45,000 K. It makes no sense that DB stars could skip this temperature range, or even move through it quickly, as white dwarfs take a long time to cool (as attested to by **ESO 439-26**). Consequently, a cooling DB star must somehow change itself into a DA star and then back again! If there is no hydrogen, where does it come from? Possibly, the non-DA white dwarfs are somewhat contaminated with hydrogen. (Hydrogen is seen in some, and though in HZ 21 some studies find it, others do not.) If so, perhaps, in this temperature range, enough is forced to the surface to hide the helium and then later dissolves again. The stars might also accumulate enough hydrogen by accretion from the interstellar medium. But nobody except the white dwarfs themselves know how the feat is accomplished, leaving the "DB gap" still unexplained.

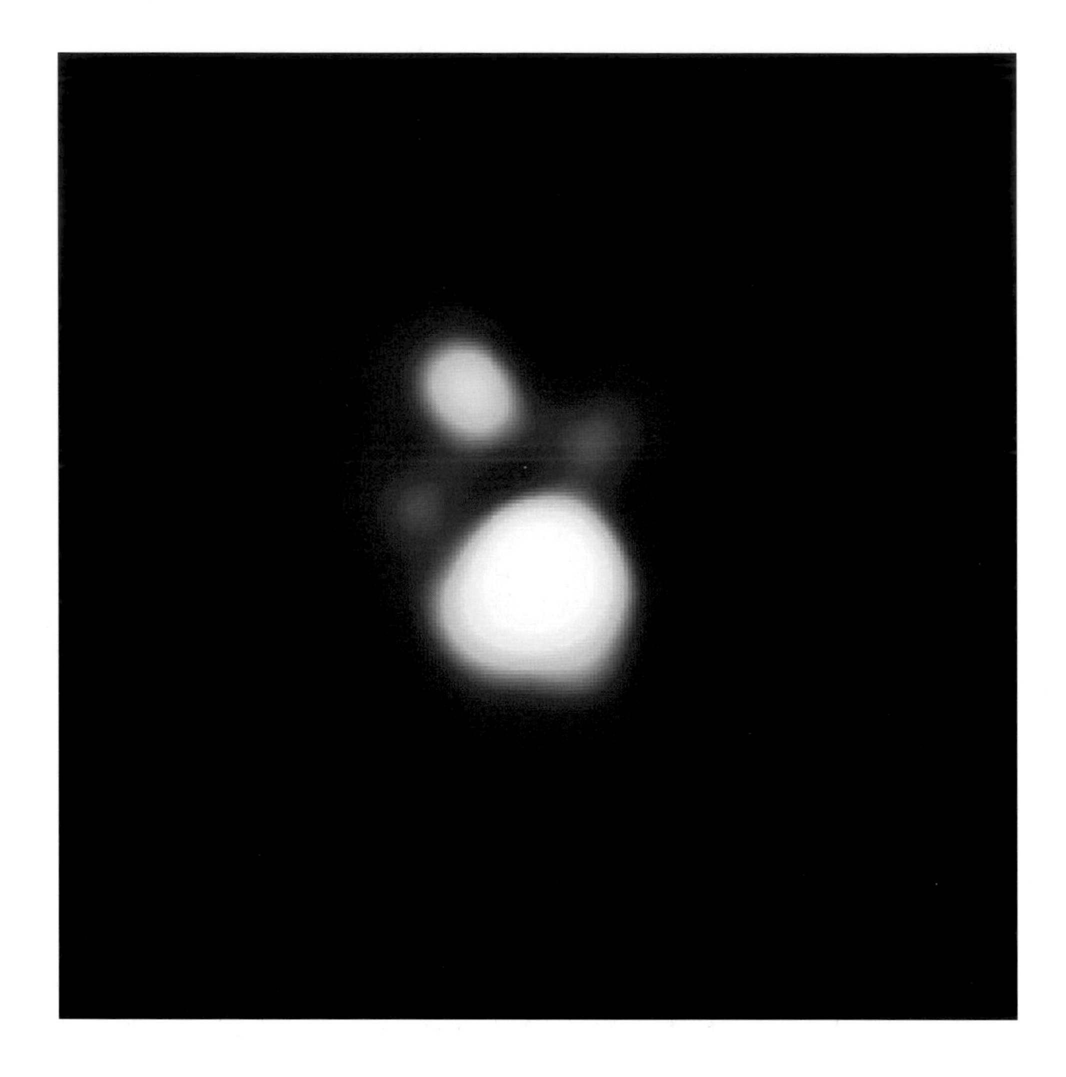

With its bipolar flow, the carbon star at the center of IRC +10 216 seems about ready to turn into a more advanced body such as the EGG NEBULA and then into a planetary nebula. *(Russian 6-meter telescope, from an article by R. Osterbart, Y. Balega, T. Blöcker, A. Men'shchikov, and G. Weigelt that appeared in* Astronomy and Astrophysics *357 [2000]: 169.)*

Residence:	LEO
Other name:	CW LEONIS
Class:	MIRA (LONG-PERIOD) VARIABLE
Visual magnitude:	NOT SEEN
Distance:	500 LIGHT YEARS
Absolute visual magnitude:	UNKNOWN
Significance:	A DUST-SHROUDED CARBON STAR.

IRC +10 216

Variables, variables, the sky abounds with variables. Among the best known is **MIRA**, the archetypal long-period variable. It is everyone's favorite "second-ascent" giant star, one with a dead carbon-oxygen core that will become a naked white dwarf (after Mira has burped its planetary nebula). Mira is famed only because at maximum it is so bright that it makes part of its constellation (Cetus); it is but one of a vast number of such variables, and not even close to any kind of limit in luminosity, temperature, or wind intensity. IRC +10 216, however, defines a kind of extreme. In spite of being about the same distance as Mira, it is effectively invisible to the eye: no one even knew that it was there until 1969, when it was found to be one of the brightest infrared sources in the sky.

IRC +10 216 is actually a huge cloud of dusty gas 1.5 light years across. At its distance of 500 light years, it stretches one tenth of a degree, 25 percent the angular diameter of the full Moon. Buried within is **CW LEONIS**, a cool Mira variable star with a temperature of about 2000 K that is so thickly covered as to be optically invisible. The star alone is extreme, with a period of variability (determined from infrared observations) of an astonishing 649 days, almost twice as long as Mira's. Unlike Mira and most long-period variables that are oxygen-rich, CW Leonis is a carbon star in the mold of **R LEPORIS**. The star's enormous luminosity of 20,000 Suns is absorbed almost entirely by the huge dusty shell and re-radiated in the infrared. Astronomers argue over the star's mass, which could be anywhere between two and five solar. Were the cloud not in the way, CW Leonis would be visually comparable to Mira in brightness.

Astronomers do not argue much about the nature of CW Leonis's shroud, however. The star is losing mass at around 0.00002 solar masses per year, and given the size and expansion velocity (15 kilometers per second), it has been doing so for close to 10,000 years, generating a cloud mass that is a good fraction that of the Sun. As the carbon-rich gas leaves the star, it chills, and some condenses to carbon dust, something akin to graphite. Theory shows that starlight pushes the dust outward. The dust, in turn, drags the gas, and the whole assembly expands steadily away.

The dust comprises about 1 percent the mass of the cloud. Most of the rest is molecular hydrogen. But remember the carbon? The element is wonderful for making molecules, rendering IRC +10 216 one of the greatest molecular factories known. Approximately 50 different molecules have been found, some not seen anywhere else. Here we find compounds from hydrogen cyanide (HCN) through ethylene (C_2H_4) and methyl cyanide (CH_3CN) to weird carbon chains (try some cyanohexatryine, HC_7N, in your coffee), even to some that might contain up to 23 carbon atoms. In some cases, carbon is not even involved. IRC +10 216 is the sole bastion of silicon dicarbide (SiC[illegible]) and of simple chlorides that include NaCl table salt!

The mass-loss process is complex. The flow rate seems to be slower than it was in the distant past, when the outer parts of the dusty cloud were being created. Moreover, it is discontinuous and may be pulsed, with thicker shells ejected every few hundred years. Deep imaging shows the formation of an inner bipolar structure that, together with the shells, is reminiscent of the structures of protoplanetary nebulae (like the **EGG NEBULA**) as well those of many full-grown planetary nebulae. It looks very much as if CW Leonis is near the end of its second-ascent giant lifetime and in the process of removing its outer layers, creating both a planetary nebula and a white dwarf.

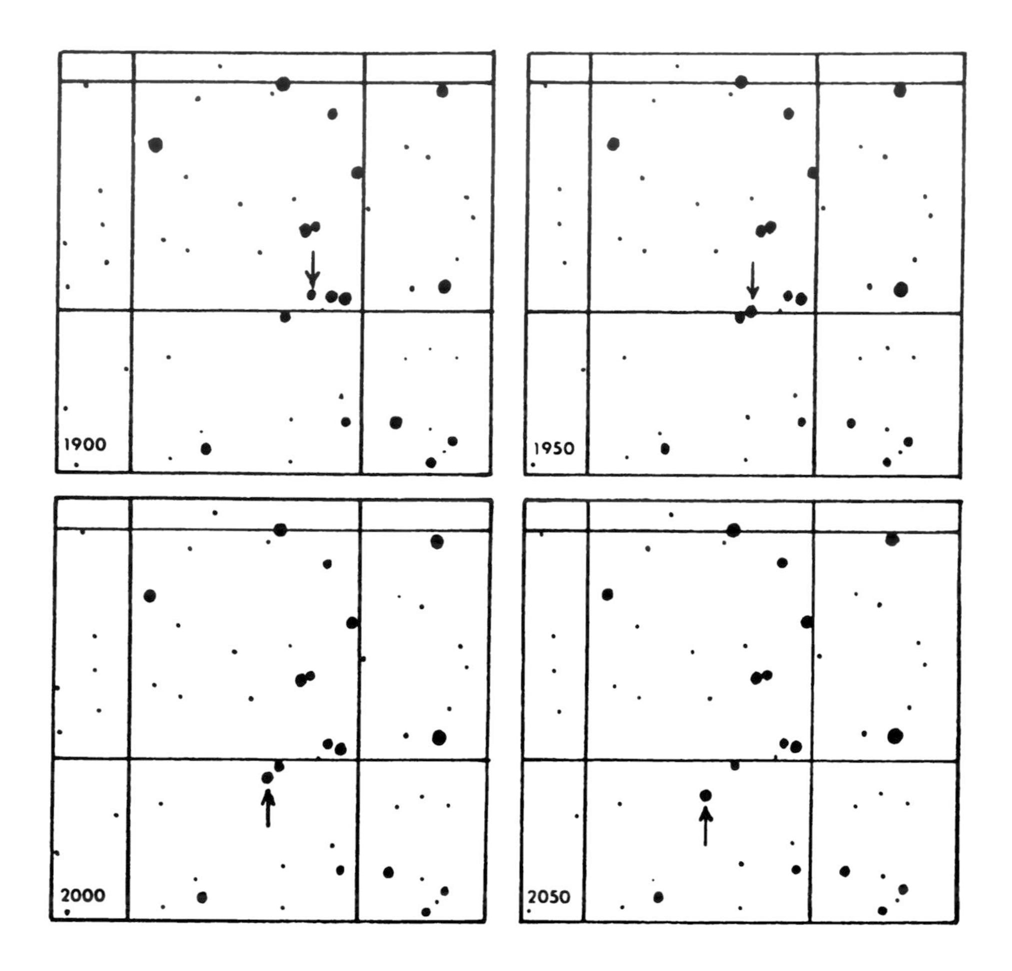

Kapteyn's Star moves quickly against the background. Each square is two degrees across, four times the angular diameter of the full moon. *(Image from R. Burnham Jr. (1978).* Burnham's Celestial Handbook. *New York: Dover.)*

Residence:	PICTOR
Other name:	CD–45°1841
Class:	M1 SUBDWARF
Visual magnitude:	8.89
Distance:	12.8 LIGHT YEARS
Absolute visual magnitude:	10.92
Significance:	HIGH-VELOCITY HALO STAR.

KAPTEYN'S STAR

All stars orbit the center of our Galaxy, which lies 26,000 light years away in the direction of Sagittarius. If they did not orbit, they would fly off into intergalactic space or fall inward to be gobbled up by the million-solar-mass black hole at the Galaxy's core. The Sun, along with 98 percent of the Galaxy's stars, belong to the plate-shaped Galactic disk and proceed around in relatively stately, circular orbits. The Sun, moving at 220 kilometers per second, takes some 225 million years to make a full circuit, a path it has traveled 20 times since birth. None of the orbits can be exactly alike, however. Small differences cause the vast majority of nearby stars to have small motions compared to the Sun, the relative speeds typically 30 to 40 kilometers per per second.

Kapteyn's Star, named after the prominent Dutch astronomer J. C. Kapteyn (1851–1922), thus presents a surprise. At first glance, it is a seemingly ordinary class M1 dwarf of the kind that abound within our local Galaxy. Among the nearest of all stars (number 24 when binary and multiple systems are counted as units), it lies a mere 13 light years away, giving an absolute visual magnitude of 10.9. After an uncertain correction for infrared radiation from a cool (3455 K) surface, the luminosity lies between a very dim 1 and 3 percent that of the Sun, and the star's mass around 0.2 solar.

But now look at the star's speed. The record-holder for "across-the-sky" proper motion is **BARNARD'S STAR**, which changes position at a rate of 10.4 seconds of arc per year, a result of a combination of closeness and high actual speed relative to the Sun. In spite of being over double the distance to Barnard's Star, Kapteyn's is a close second, zipping along at 8.7 seconds of arc per year (the angular diameter of the full moon in two centuries). Now in the constellation Pictor, Kapteyn's Star is headed southwesterly toward Dorado, which it will pass into in about 3500 years. To have such a motion, the star must be moving across the line of sight at 160 kilometers per second, a speed 50 percent greater than that of Barnard's. To that, we add a spectacular radial velocity (the speed directly at us) of 245 kilometers per second. Together, the two speed components give Kapteyn's Star a record-breaking (for nearby objects) and breathtaking velocity of 294 kilometers per second relative to the Sun.

Such a speed could only be generated by a Galactic orbit that is very different from our solar path. Kapteyn's Star is not of our Galaxy's disk, but of the halo. On a long elliptical orbit, Kapteyn's Star is hurtling downward through the Galaxy's plane, traveling across the solar path. The situation is rather like walking down Fifth Avenue in Manhattan and having somebody who is running along 54th Street cross in front of you.

The Galaxy's halo is older than its disk, and the stars in it were created before stellar evolution could build up a heavy load of metals (everything other than hydrogen and helium). As a visitor from the halo, Kapteyn's Star is notably low in metals, with a metal content that has been estimated to be as low as 1 percent that of the Sun. Lack of metals makes a stellar atmosphere more transparent and changes the star's radius and temperature. As a result, for a given luminosity, stars such as Kapteyn's commonly appear hotter than ordinary or, for the same temperature, less luminous, placing them into the dismal subdwarf category, not a very nice thing to say about such an informative, though hurried, guest.

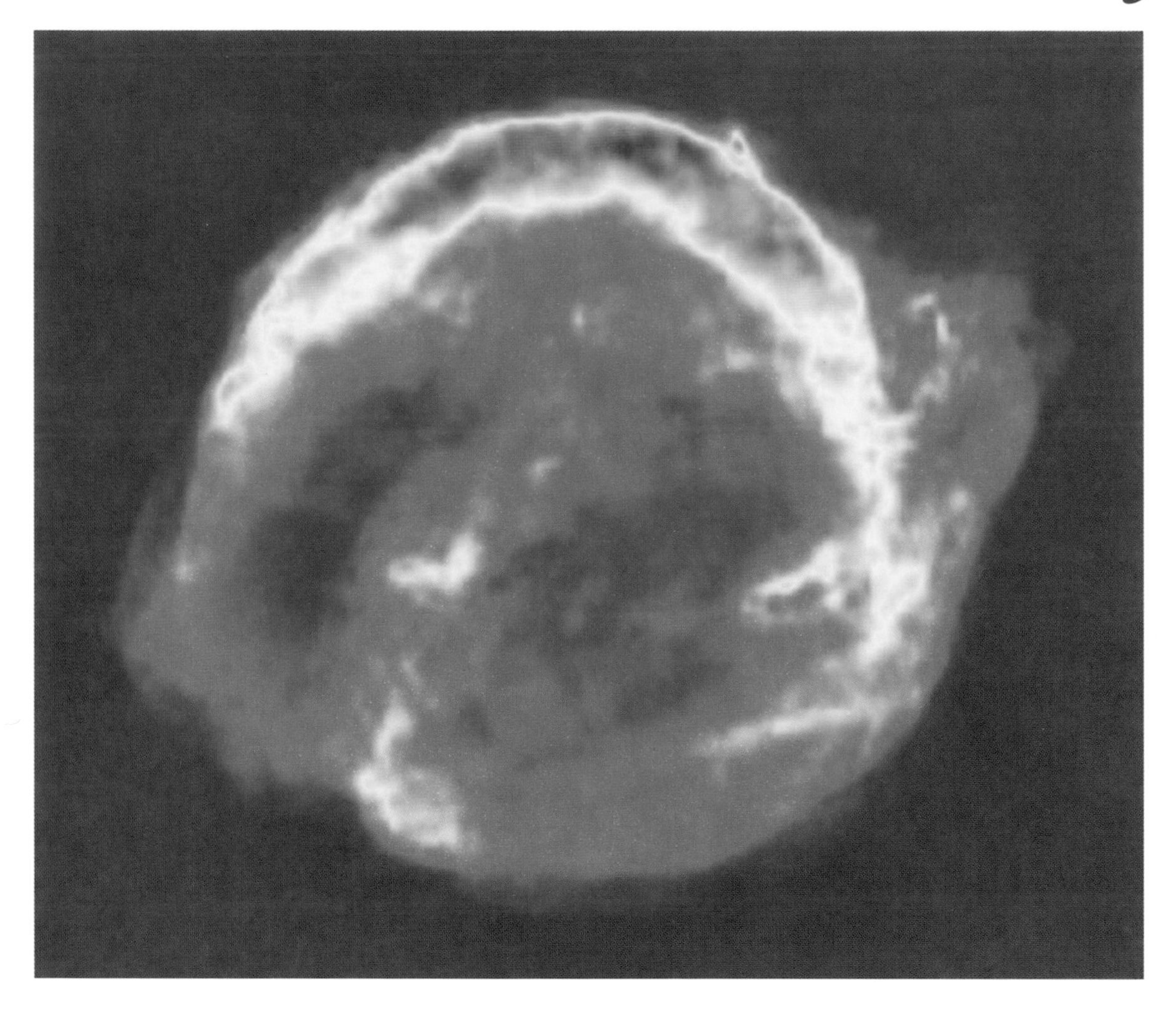

Although the expanding remnant of Kepler's Star (now 15 light years across) is optically almost invisible, it radiates fiercely in the radio spectrum as a result of the shock wave of the explosion hitting interstellar matter. *(Very Large Array, National Radio Astronomy Observatory, T. DeLaney, B. Koralesky, L. Rudnick, and J. R. Dickel, from a paper in the* Astrophysical Journal *2002, in press.)*

Residence:	OPHIUCHUS
Other name:	V 843 OPHIUCHI
Class:	SUPERNOVA
Visual magnitude:	AT MAXIMUM –2.2
Distance:	18,000 LIGHT YEARS
Absolute visual magnitude:	AT MAXIMUM –18.9
Significance:	THE LAST SUPERNOVA SEEN IN OUR GALAXY.

KEPLER'S STAR

We know that the brilliant star discovered in 1604 by Johannes Kepler, who carefully recorded its fading, was a supernova. The last one seen in our Galaxy, it blew up before we had telescopes. However, by observing supernovae in other galaxies, we have learned that they come in two distinctive flavors. The spectra of Type II supernovae show hydrogen absorptions; they are found only in the disks of galaxies where massive stars reside. Type I are hydrogen-less and can be found anywhere, in the disks to be sure, but also in surrounding Galactic halos and in elliptical galaxies, which are dominated by low-mass stars. Stars in general, however, are divided into two "populations," those of a galaxy's metal-rich disk Population I, those found in metal-poor halos Population II. Type I is Pop. II, Type II is Pop. I.

A Type II supernova is caused by the collapse of the iron core of a massive star (which the star built from the original hydrogen). Type I is further divided into Ia, which shows strong silicon absorption, and Ib, which does not. It is the Ia supernovae that appear in galaxy halos. Like **TYCHO'S STAR**, they are created when white dwarfs in double stars somehow step over the allowed limit of 1.4 solar masses and collapse. Ib supernovae are probably produced by the collapse of massive Wolf-Rayet stars like the one in **GAMMA-2 VELORUM**.

Kepler's Star was found in the autumn, deep in southern Ophiuchus. Reaching magnitude –2, about the brightness of Jupiter, it was quickly overtaken by the Sun. By the time it was recovered in the morning sky, it had dimmed to first magnitude. A bit over a year after its discovery, Kepler's Star dropped below naked-eye visibility and was lost forever. In 1943, Walter Baade reconstructed the "light curve" (the record of magnitude versus time) from Kepler's and others' observations, postulating that it was not an ordinary nova like **DQ HERCULIS**, but a supernova.

All that remains of the event is the expanding debris, consisting of a few shards of optically visible gas, strong radio emission, and powerful X-rays from gas heated by shock waves. X-ray and radio images show a circular, shell-like structure 3.3 minutes of arc across, expanding at a speed of 5000 kilometers per second. The distance is very uncertain: the best estimate is 18,000 light years, which makes the remains 16 light years across. Factoring in interstellar dust absorption, that distance gives a maximum absolute visual magnitude of –18.9 (3 billion times the solar visual luminosity), right on target for a supernova. If it had gone off at **VEGA**'s distance, it would have lit up the sky with the brightness of 500 full moons.

Because spectroscopy in 1604 was limited to rainbows, we still do not know quite how to classify Kepler's Star. Its location of 2000 light years above the Galactic disk strongly implies that it was Type Ia. However, the optically visible remnants expand away slowly, suggesting that they are not the debris of the explosion. Instead, they appear to be part of a huge gaseous shell that had enveloped the star prior to the detonation, something expected from a massive star that had had a strong wind. Moreover, X-ray observations show only a small amount of iron created in the explosion, appropriate for a Type II or Ib supernova, but not for a Ia which typically makes about half a solar mass of the stuff.

How could a massive star get that far from the Galaxy's plane? Gravitational ejection from a cluster, or a kick from a companion that also "went supernova," could have made the progenitor a runaway star like **MU COLUMBAE**. Nearly four centuries after its discovery, Kepler's Star is still wrapped in mystery.

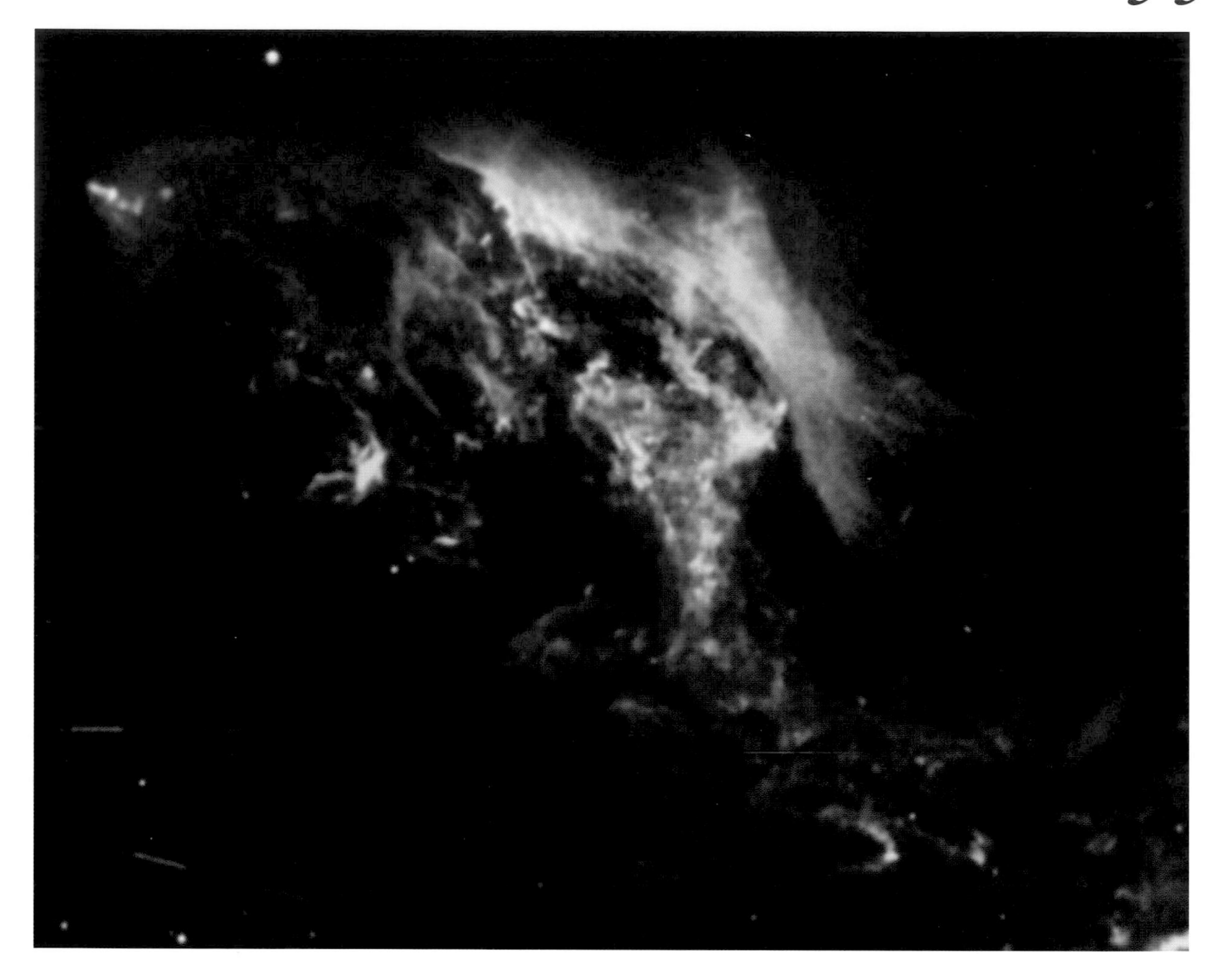

A complex set of gas flows and shock waves are involved with the region that is forming the young star L 1551 IRS 5, which is buried in the cloud toward the upper left. *(ESO New Technology Telescope, courtesy of B. Reipurth and J. Bally.)*

Residence:	TAURUS
Other name:	IRAS 04287+1801
Class:	YOUNG STELLAR OBJECT
Visual magnitude:	NOT KNOWN
Distance:	450 LIGHT YEARS
Absolute visual magnitude:	AN UNCERTAIN 1.0
Significance:	A STAR—OR STARS—IN THE PROCESS OF BIRTH.

L 1551 IRS 5

Dark clouds easily visible to the naked eye splatter the Milky Way. They are interstellar clouds many light years wide and so thick with dust that they block the background light. Nearly 100 years ago, E. E. Barnard suggested that the dark clouds are the birthplaces of stars. No one could see stars being born, so the process must be hidden, and where best but by the obvious dark blobs? Thin, transparent interstellar clouds are heated to hundreds, even thousands, of degrees by starlight. However, the obscuring dust in the clouds makes them so cold that they become immense molecule factories. Portions of the cold molecular gas are forced to contract under their own weight until stars are formed within.

Direct evidence came from youthful stars like **T TAURI** and **FU ORIONIS** that flock around the dark clouds, apparently having escaped their innards. More evidence accumulated in the form of Herbig-Haro objects, small nebulae discovered by George Herbig and Guillermo Haro in the dark-cloud regions that seemed to have no sources of illumination. Then, lo!, the "HH objects" were found to come in pairs with T Tauri or related stars between them. The HH objects are created by oppositely directed, bipolar jets that flow from the new stars. The jets sweep up the surrounding interstellar gas, creating a "bow shock" similar to the shock wave that flies from the bow of a speedboat. Along with other evidence, the HH objects and bipolar jets suggest that the young stars are surrounded by jet-directing disks that lie perpendicular to the flows: disks from which planets might be born. What precedes these young things?

The first comprehensive list of photographically observed dark clouds was created by Barnard himself. It was replaced in the mid-twentieth century by B. T. Lynds's catalogue. Number 1551 in the Lynds Catalogue lies in the heart of Taurus, in the back of the famed Hyades cluster (which makes the Bull's head) and about three cluster-distances away. A pair of very young T Tauri stars are associated with Lynds 1551, as is a reflection nebula (somewhat like that around the Pleiades' **MEROPE**) and numerous HH objects. Buried within are several infrared sources (abbreviated "IRS"). We zero in on the one the catalogues label "5," hence L 1551 IRS 5.

Emerging from IRS 5 is a spectacular flow of molecular gas outlined by radio radiation that is produced by easily-observed carbon monoxide and is moving at some 200 kilometers per second. Surrounding the object is a very dense disk illuminated by radio radiation from carbon monosulfide. A pair of small, narrow jets also emerges from the cloud. The actual object is impossible to see in optical radiation as a result of 30 to 100 magnitudes of absorption produced by the surrounding dust. High-resolution radio observations, however, reveal a pair of sources separated by just 0.35 seconds of arc, or about 50 AU, just a bit farther than Pluto orbits from the Sun. We appear to be seeing the actual birth of a double star. The total luminosity, derived from infrared observations, is around 30 times that of the Sun and implies a pair of ordinary stars of modest (but notably greater than solar) mass.

Stars begin as huge, slowly rotating collections of gas and dust that have been sufficiently compressed, perhaps by supernova blasts. As they contract, they rotate faster and faster. They are kept from flying apart by a brake applied by the Galaxy's magnetic field, by the flows that can carry rotational energy away, and by splitting into pairs to produce double stars. Where in the overall scheme of star formation L 1551 IRS 5 actually falls is not very well known, but it is a "young stellar object," if ever there was one.

Boötes's principal treasure is orange ARCTURUS (seen at right). Toward the upper left and circled is another, the oddly metal-deficient star Lambda Boötis. *(J. B. Kaler.)*

Residence:	BOÖTES
Other name:	19 BOÖTIS
Class:	A0 PECULIAR DWARF
Visual magnitude:	4.18
Distance:	97 LIGHT YEARS
Absolute visual magnitude:	1.81
Significance:	ODDLY METAL-DEPLETED STAR.

LAMBDA BOÖTIS

Nineteenth-century astronomers found the spectra of stars so varied that it seemed obvious that the stars had to be made of different substances. After we came to understand how spectra were produced, we learned that ordinary dwarfs all have pretty much the same chemical compositions. Ignoring their nuclear-burning cores, dwarfs are 90 percent hydrogen, 10 percent helium, and about 0.2 percent everything else; the remainder includes oxygen, carbon, neon, nitrogen, and so on in solar proportions. Differences among dwarf spectra and the spectral sequence are caused by temperature differences alone.

But not so fast. As it turns it out, there are a great number of different actors on the stage. First up are the metal-poor halo stars, represented *in extremis* by **CD–38°245**, even by local subdwarfs like **KAPTEYN'S STAR**. Class A stars like **CHI LUPI** show peculiar abundances as a result of diffusion of elements, and others, like **COR CAROLI**, display the effects of magnetic fields.

And then there is Lambda Boötis. From its distance of 97 light years, Lambda glows with a luminosity 16 times that of the Sun. Its temperature of 8700 to 8900 K, somewhat too cool for the spectral class, provides the first hint that something is a bit awry. Together, temperature and luminosity tell of a nearly two-solar-mass star that rotates at least 65 times more rapidly than the Sun, spinning in 16 hours or less.

Spectra reveal that Lambda Boo has anomalously faint metal absorptions (here "metal" used in the true sense of iron and the like). The star is the leader of a small group of about 50 known similar stars whose metal abundances (which include chromium, barium, nickel, titanium, even non-metal silicon) are reduced by about a factor of 10, while things like oxygen and carbon are in more or less normal proportions. Their motions, along with the normal oxygen and carbon abundances, reveal that Lambda Boo stars are not from the low-metal halo. So, at first, they appear as one more set of chemically peculiar stars (again like Chi Lupi), whose abundances are altered by diffusion processes that pull some kinds of atoms downward under the force of gravity, while others are lofted by radiation. However, the rapid rotation of the Lambda Boo stars should keep their atmospheres stirred up and the chemical elements mixed in their proper proportions. The weird abundance patterns suggest that gas is being accreted from interstellar space.

Interstellar dust is produced by stars in advanced states of evolution like **MIRA** and other massive mass-losing stars. In the cold of space, the dust grains slowly accumulate individual atoms from the gas as ices—but only those that freeze onto the particles at relatively high temperatures. Oxygen and its cohort are too volatile to join the solid brigade, but metal atoms are generally not. As a result, metals can be relatively depleted from the gas by huge factors, leaving them factors of hundreds short of their normal abundances.

What happens if an ordinary star is heavily surrounded by a cloud of enriched dust and depleted gas? The gas will be accumulated by the star, but the dust grains will be blown outward by the pressure starlight, resulting in a metal-poor stellar skin. Combined with some up-and-down separation of elements, this theory seems to explain the Lambda Boo stars quite well. And yet the mixing of stellar gases caused by the rotation should destroy the depletions rather quickly. So the stars must be young or still accreting, and the evidence is mixed at best. Lambda Boo is trying to tell us something. We just do not know quite what it is.

LB 11146, the tiny star at the center of the picture (which is a half-degree across), harbors a double white dwarf. (© National Geographic—*Palomar Observatory Sky Survey, reproduced by permission of the California Institute of Technology.)*

Residence:	LEO
Other name:	WD 0945+245
Class:	DOUBLE WHITE DWARF
Visual magnitude:	14.32 (COMBINED)
Distance:	130 LIGHT YEARS
Absolute visual magnitude:	11.29 (COMBINED)
Significance:	SUPPORTER OF A SUPERNOVA THEORY.

LB 11146

Double stars abound in space. Being in a double system is almost more normal for a star than being single. Of the nearest 70 stars (which may not be typical), about half are in doubles or multiples, and many of the apparently single stars may well have yet-undetected companions. Since almost all stars, with the exception of exploders above ten solar masses, evolve to white dwarfs, double white dwarfs ought to abound. They are enormously important, because they may be *en route* to becoming Type Ia supernovae, like those found in galaxy halos (and like **TYCHO'S STAR**). Yet only a few dozen are known.

Most Type Ia supernovae reach absolute magnitude –19 at their peaks. As a result, they have become prime "standard candles" for measuring the distances to very remote galaxies and studying the large-scale expansion of the Universe. We are nearly certain that the Ia supernova process involves pushing a white dwarf over the Chandrasekhar cliff, beyond the 1.4-solar-mass white-dwarf limit, above which a white dwarf cannot support itself.

Two competing theories attempt to account for the production of white-dwarf supernovae. In the more popular scenario, a white dwarf closely circles a low-mass main sequence star. The two have been drawn together by friction while wrapped in common envelopes produced by the earlier giant-star stages of evolution. Tides raised by the white dwarf in the ordinary dwarf allow the flow of mass from the bigger star to the smaller. When enough fresh hydrogen has accumulated, the white dwarf's surface may blow up to produce a nova like **DQ HERCULIS**. But if the white dwarf is already near the allowed mass limit, it might be eased beyond it before the nova can occur. Then, the white dwarf will violently collapse and explode. In the second scenario, two white dwarfs closely circle each other. Orbital decay (brought about by gravitational radiation predicted by relativity) may bring them close enough to merge, and if the sum of masses is greater than 1.4 solar, bang! And yet no white dwarfs with sufficiently high combined masses were observed.

That is, not until LB 11146 was found. The star's spectrum can be replicated theoretically only by assuming the presence of two white dwarfs with slightly different temperatures near 15,000 K. The individual parameters are remarkably similar, making them look like descendants of double class A stars rather like those in **EPSILON LYRAE**. Each one contains about 90 percent the mass of the Sun tucked within a radius about 95 percent that of Earth. From their observed magnitudes and average luminosities of 0.004 solar, we work backwards to find a distance of 130 light years, which for such faint stars is beyond easy and accurate direct measure.

But now the similarity between the two stars comes to an abrupt halt. Component B, like **EG 129**, is highly magnetic, with a field over 300 million times stronger than Earth's, while "A" is not. Since magnetic white dwarfs probably descend from magnetic class A or B dwarfs, the original system was probably a peculiar (magnetic) A star like **COR CAROLI** coupled with a normal A star. Why such systems exist, however, is still a mystery.

The separation of the LB 11146 pair is unknown, but Hubble Space Telescope observations show the two to be less than 2.3 AU apart, which would yield an orbital period of under 2.6 years. At such a distance, they would not be able to merge. Yet the very existence of such a system, whose total mass exceeds the famed white-dwarf limit, gives some credence to the minority view of how white-dwarf supernovae can come about.

The Pleiades star cluster is passing through a wispy interstellar cloud of gas and dust, the latter scattering blue starlight. The nebula is especially bright near Alcyone, toward the top, and Merope, near bottom center. *(NOAO/AURA/NSF.)*

Residence:	TAURUS
Other name:	23 TAURI
Class:	B6 SUBGIANT
Visual magnitude:	4.18
Distance:	385 LIGHT YEARS
Absolute visual magnitude:	–1.39
Significance:	BRIGHT PLEIADES STAR AND ILLUMINATOR OF THE MEROPE NEBULA.

MEROPE

Little in the sky is more attractive than the delightful Seven Sisters, the Pleiades of Taurus, which make a bright compact "open cluster" filled with blue-white class B gems. Seven of the stars in order of brightness, Alcyone, Electra, Maia, Merope, Taygeta, Celaeno, and Sterope are named for the sisters who, in Greek mythology, were the daughters of the god Atlas and mortal Pleione, who appear as well. If all nine are included, the parents respectively rank 2 and 7. Of the Pleiades, only Alcyone carries a Greek letter name (Eta); the rest are designated by Flamsteed numbers, which count naked-eye stars within a constellation from west to east. As a result, **MEROPE** is also called 23 Tauri.

Along with the rest of the cluster, Merope, a B6 subgiant, lies at a well-determined distance of 385 light years. From its 14,000-K surface, the star shines with a total luminosity of 630 times that of the Sun. This Pleiad is a particularly fast rotator, spinning with an equatorial speed of at least 280 kilometers per second, 140 times that of the Sun. Given its radius of 4.3 solar, the star makes a full rotation every 18 hours or less. The fast rotation affects the star's immediate environs and spectrum. Like three others in the Pleiades (Alcyone, Electra, and Pleione), Merope is an emission-line "Be" star, its rapid spin flinging out a disk of bright gas, though Merope's is quite thin. Nevertheless, the disk is sufficiently dense and hot from shock waves to produce observable X-rays.

Merope's other claim to fame is its extended surroundings. The Pleiades are enmeshed in a cloud of dusty gas. But the stars are not hot enough to ionize the gas and make it glow as **THETA-1 ORIONIS C** does in the Orion Nebula. Instead, the tiny dust grains embedded in the cloud scatter and reflect the starlight to make the Pleiades Reflection Nebula. Since solid grains scatter short-wave light (violet and blue) more efficiently than long-wave yellow and red, and since the stars are blue-white to start with, the nebula glows a lovely blue.

The reflection nebula is at its best around Merope, where it is bright enough to have special names: the "Merope Nebula" and "IC 349." ("IC" stands for "Index Catalogue," a large addendum to the standard nineteenth century "NGC," or "New General Catalogue" of celestial objects, which, in turn, is an updating of the great Herschel "General Catalogue." The NGC and IC together contain over 13,000 clusters, nebulae, and galaxies.) Long thought to be the remnant of the Pleiades birth, the nebula is instead a chance encounter, as the cluster is merely passing through a random interstellar cloud. The Pleiades actually leave a wake as the cloud rushes past.

The bright named stars of the Pleiades are all relatively massive, and all but two are either giants or subgiants, that is, they have evolved off the main sequence, or are at least beginning to. Merope, a subgiant, contains some 4.5 solar masses; only Pleione and Sterope are classed as dwarfs. Since higher-mass stars die first, we can date the cluster using the theoretically calculated age of the most massive dwarf stars left behind as the others trek on to become giants. (In practice, astronomers theoretically replicate the array of luminosities and temperatures of all the stars in the cluster.) Merope and the other Pleiades can be no more than 100 million years old. Bright and beautiful, the Sisters are still Galactic children.

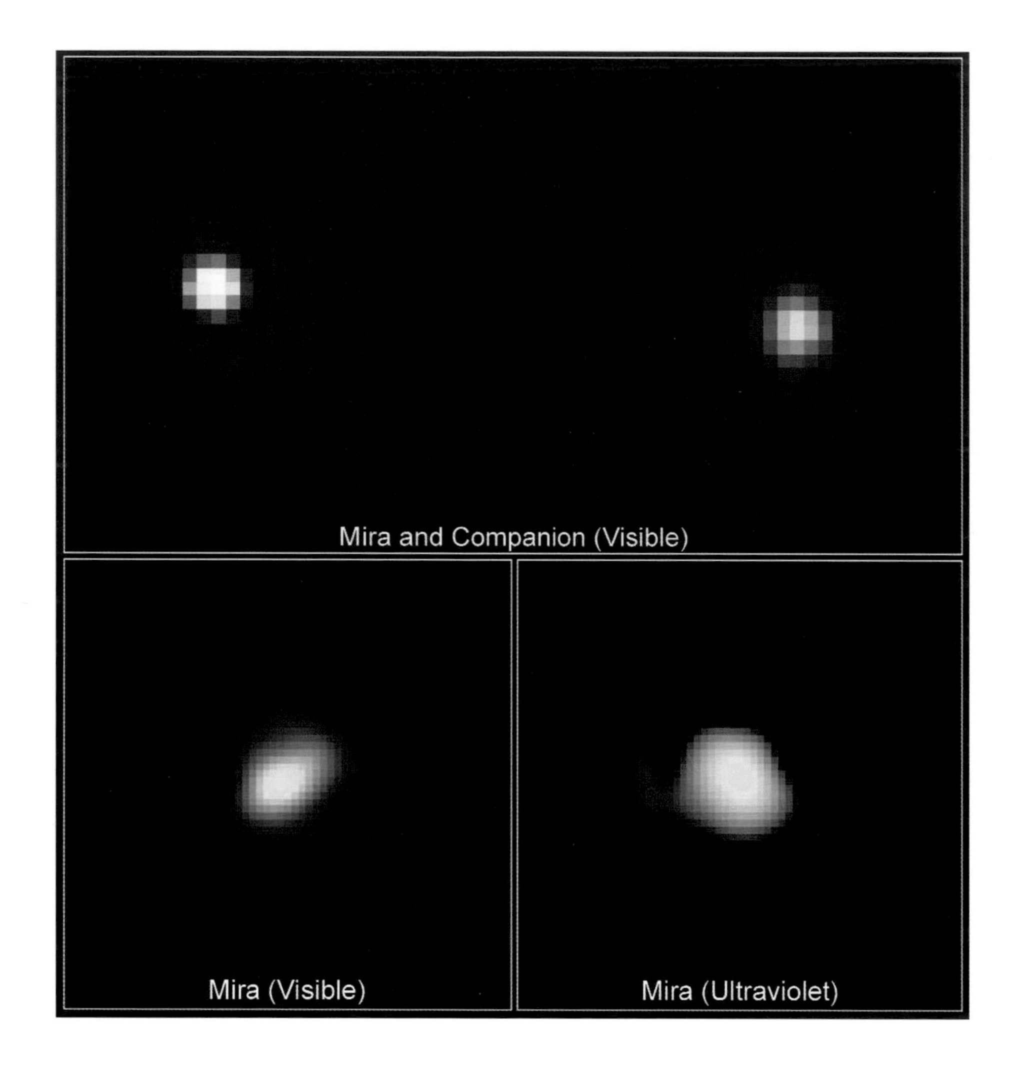

Mira, imaged by the Hubble Space Telescope, is not round, perhaps as a result of the influence of its white-dwarf companion. *(Hubble Space Telescope, M. Karovska [Center for Astrophysics], STScI, and NASA.)*

Residence:	CETUS
Other name:	OMICRON CETI
Class:	M7 (AVERAGE) GIANT MIRA
Visual magnitude:	3 TO 10
Distance:	420 LIGHT YEARS
Absolute visual magnitude:	–3 TO –5
Significance:	THE GRAND, LONG-PERIOD VARIABLE AFTER WHICH ALL THE OTHERS WERE NAMED.

MIRA

What is the sky's greatest star? **POLARIS** at the Pole? Brilliant **SIRIUS**? **TYCHO'S STAR**? Perhaps it would be Mira, the very name proclaiming glory: the "Miracle Star," "the Amazing One," Mira "the Wonderful." It was found in 1596 as a "new" third magnitude star in Cetus by David Fabricius, a friend of Tycho Brahe. Like Tycho's luminous star, it disappeared. The wonder of Mira is that, unlike Tycho's, it came back! And then disappeared again. It returns about once a year and has made some 450 appearances since its discovery. And we can rely on it coming back a hundred thousand times more (providing us with even more shows than *Cats*).

Mira typically varies between apparent magnitudes 3 and 10 (a factor of 600) over a period of 330 days, but it can be as bright as first magnitude. The star is a giant of vast proportions. Even at 420 light years away, it is easily detectable as a disk 0.048 seconds of arc across, from which we derive a physical radius of 3 AU (double of the size of the Martian orbit). Mira's visual luminosity changes from 1 to nearly 10,000 solar. The star is the prototype of the "Mira variables," of which thousands are known.

Mira is also the archetype of the "second-ascent" giants, of the awkwardly-named "asymptotic giant branch," or "AGB," stars. Mira is in this league with variables such as **CHI CYGNI** and **R LEPORIS**. Having left the main sequence long ago, such stars contain contracting carbon-oxygen cores nested in burning shells of helium and hydrogen. As these second-ascent stars slowly grow more luminous, they become unstable and begin to pulsate, forcing their radii, spectral classes, and surface temperatures to fluctuate. At Mira's hottest (3000 K), it radiates a fair degree of visual radiation. When the temperature drops into the 2000-K range, the light emerges much more in the infrared, and the star dims. The fading is enourmously helped by the formation of titanium oxide molecules that fiercely absorb visual radiation. The star thus fools the eye, as the total variation in luminosity is only about two magnitudes (a factor of 6), far less than we see with our eyes. Including its infrared radiation, Mira blasts 7000 solar luminosities into space.

Mira is so large that its apparent disk has even been imaged. Instead of being round, the star looks more like a distorted water balloon, a behavior that may be internally generated. It may also be driven by the gravitational action of a faint white-dwarf companion named Mira B. Whatever the reasons, the pulsations drive powerful shock waves through the star's outer atmosphere, propelling a wind in which some of the gas condenses into dust grains. Most of the star-forming dust of interstellar space comes from such wondrous giants.

Mira is losing mass at a modest rate of a bit under a millionth of a solar mass per year (over 1 million times the flow rate of the solar wind). As it sheds its outer envelope, it has become enmeshed in a huge molecular shell. Observations of radio radiation from carbon monoxide show us that the shell stretches some 200 Mira-radii from the stellar surface. The wind appears to interact with the white-dwarf companion, making Mira a "mild symbiotic star," a lesser version of more dramatic symbiotics like Chi Cygni or **R AQUARII**.

As Mira loses its outer envelope, it will slowly reveal its nuclear furnace, a white-dwarf-to-be. As the star evolves, the hot core will for a time compress and light the fleeing mass, and a new planetary nebula (another **NGC 6543** perhaps) will grace the sky, our Wonderful Star re-emerging from a cocoon to produce a celestial butterfly and a double white dwarf.

The Big Dipper stands vertically above the distant lights of Tucson, Arizona. The end of the handle is lost in the thick atmosphere above the horizon, but the next star, Mizar, with its companion Alcor, is nicely seen. *(J. B. Kaler.)*

Residence:	URSA MAJOR
Other name:	ZETA URSAE MAJORIS (ALCOR: 80 URSAE MAJORIS)
Class:	A1/A7 QUARTET (ALCOR: A5)
Visual magnitude:	2.06 (ALCOR: 4.01)
Distance:	83 LIGHT YEARS
Absolute visual magnitude:	0.99/0.99/2.68/2.68 (ALCOR: 2.03)
Significance:	NAKED-EYE DOUBLE, THE ORIGINAL TELESCOPIC DOUBLE, AND A FINE QUINTET

MIZAR (AND ALCOR)

You would think that one of the best-known stars in the sky would be completely understood. Think again. The most famed of "double stars" is Ursa Major's naked-eye pair Mizar and Alcor, the only double for which each member has a proper name. But is it really double? We are still not sure. "Mizar," derived from the Arabic word meaning the "groin" of the Bear, and Alcor, from the same word for "bull" that was applied to the neighboring Dipper star Alioth, make the Arabs' "Horse and Rider." Two-tenths of a degree apart, respectively second and fourth magnitudes, they provide a test of minimal vision.

In 1650, Mizar achieved fame as the first telescopic double star. A low-power eyepiece reveals a pair of white class A stars 14 seconds of arc apart. The spectrograph then showed each one of these to be double, making it a variation on the **EPSILON LYRAE** theme. The brighter of the visible pair, second magnitude Mizar A, consists of two class A1 dwarfs with temperatures of 9000 K in a tight orbit with a period of only 20.5 days. Remarkably, though the stars are typically but 0.4 AU apart, they have been separated by interferometer. These devices observe the way the light interferes with itself, much in the way the waves spreading from two rocks dropped into a pond interact. Interferometers are used to measure stellar diameters and the separations of close doubles. The observed orbit and Kepler's laws tell of two stars, each with masses 2.5 times that of the Sun. The pair that makes fourth magnitude Mizar B are about 1 AU apart. Classed somewhere between A1 and A7, with temperatures of 8000 K and masses 1.5 solar, they make their orbital turn in half a year. The doubles themselves are at least 500 AU apart and take more than 5000 years to circulate.

Like so many stars of classes F, A, and B, each of the pairs has strange chemical compositions caused (like that of **CHI LUPI**) by the physical separation of elements in quiet atmospheres. Slow rotation keeps the stars from stirring themselves. The Mizar A double is a "peculiar" star, heavy on strontium and silicon, whereas Mizar B is a "metallic line" star that is quite deficient in aluminum and calcium but high in silicon and in rare-earth elements like cerium and samarium. Alcor, on the other hand, seems to be single. A class A5 dwarf a bit hotter than the Mizar B stars, it is a slightly pulsating variable. Rotating rapidly, it mixes its atmosphere and presents normal chemical abundances.

Now, are Mizar and Alcor coupled? It is still hard to say. Precision parallaxes with the Hipparcos satellite show Mizar to be 78.1 light years away, while Alcor is 81.1 light years distant. Mizar and Alcor are part of the Ursa Major cluster, whose core consists of the middle five stars of the Big Dipper. A separation of over three light years, almost the distance between here and Alpha Centauri, would make a gravitational pairing unlikely because the gravity of neighboring stars would pull them apart. The measured errors allow a separation as close as 0.7 light years, but even at this distance, Mizar and Alcor would orbit at only 0.4 kilometers per second, and the measured speed difference is two and a half times higher. The errors in the Hipparcos distances are suspected of being greater than listed, however, and the analysis of the orbit of Mizar A suggests that Mizar might actually be farther than Alcor! If they are actually at the same distance, their minimum separation is only 0.27 light years, making them close enough so that they could truly orbit, though with a long period of 0.75 million years. Conclusion? They are probably paired, Alcor, the "rider," firmly saddled upon Mizar, its four-legged steed.

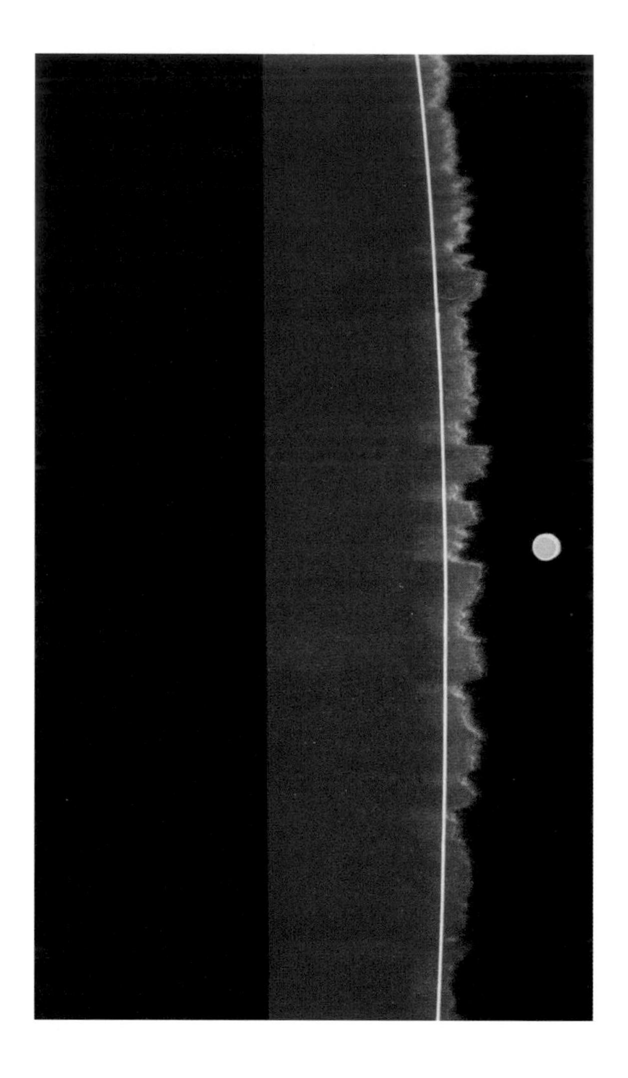

Huge Mu Cephei dwarfs the Sun, indicated by the small circle to the right. *(From J. B. Kaler, 1994.* Astronomy! *New York: HarperCollins.)*

Residence:	CEPHEUS
Other name:	HERSCHEL'S GARNET STAR
Class:	M2 BRIGHT SUPERGIANT
Visual magnitude:	4.1 (VARIABLE)
Distance:	2000 LIGHT YEARS
Absolute visual magnitude:	–7.3
Significance:	AMONG THE LARGEST AND BRIGHTEST STARS IN THE GALAXY.

MU CEPHEI

Those who think that stars are not highly colored need only look at Mu Cephei, often called "Herschel's Garnet Star" (honoring both this cool, red star and William Herschel, who discovered infrared radiation). The color of this fourth magnitude star, easily found in southern Cepheus, will leap out at you.

Mu Cephei's wonderful color is only the beginning. Classed as an M2 Ia supergiant, the "Ia" implying the brightest kind, it is among the most luminous and largest stars in the Galaxy. The distance is rather uncertain. Too far away for a meaningful measure of individual parallax, the distance is found through membership in a stellar club. A large fraction of hot, massive blue class O and B stars belong to loosely bound "OB associations," a group of stars more or less born together. When hydrogen-to-helium fusion ceases in a massive O star, it expands and cools to become a red supergiant that fuses its now-helium core to carbon and oxygen. As a result, we find red supergiant rubies set upon necklaces of blue O- and B-star diamonds. The Garnet Star belongs to the Cepheus OB2 association, and from the combined, statistically valid distances of the other members, we find it to be about 2000 light years away.

The distance is but one of three factors in determining Mu Cephei's luminosity and size. A star this distant and located in the Milky Way will be subject to considerable dimming by interstellar dust. Were no dust present, Mu Cephei would be 2.5 magnitudes brighter, near first magnitude, causing dim Cepheus to appear far more prominent. The low temperature implied by the star's redness and spectrum (around 3600 K) tells us that much of the radiation emerges in Herschel's invisible infrared. When these corrections are made, the Garnet Star is seen to radiate 350,000 times more energy than the Sun, which, in turn, implies a radius of 7 AU—40 percent larger than the orbit of Jupiter! Mu Cephei is so large that its actual apparent disk is readily discernable. Its measured angular diameter of 0.018 seconds of arc leads to a direct determination of a radius of 5.5 AU, a bit smaller but still huge. The difference between the two measures reflects the uncertainties in distance, dust absorption, temperature, infrared correction, and the degree to which the star has a fuzzy apparent "edge" (it may be angularly larger than measured).

If such numbers seem hard to believe, look just a bit north to nearly identical **VV CEPHEI**, which belongs to the same association. Every 20 years the supergiant eclipses a bright class O companion. The duration of the eclipse shows VV Cephei to be nearly 9 AU in radius, strongly suggesting that the higher value of Mu's radius is correct. Mu Cephei's mass must be in the neighborhood of 40 times that of the Sun. It will surely explode someday.

Like all stars of its size and luminosity, the Garnet Star is unstable and is losing mass. A semi-regular pulsating variable, it wobbles in brightness from nearly third magnitude to fifth with ill-determined periods of 2.3 years and 12 years. Its wind, now blowing at a rate of about a tenth of a millionth of a solar mass per year (100,000 times greater than the solar wind), has created a huge surrounding shell that is a third of a light year across. Loaded with dust, the shell is the source of laser-like silicon monoxide emission, while the extreme inner portion contains hot water vapor. Best of all, unlike so many extreme stars, you can see it with the naked eye.

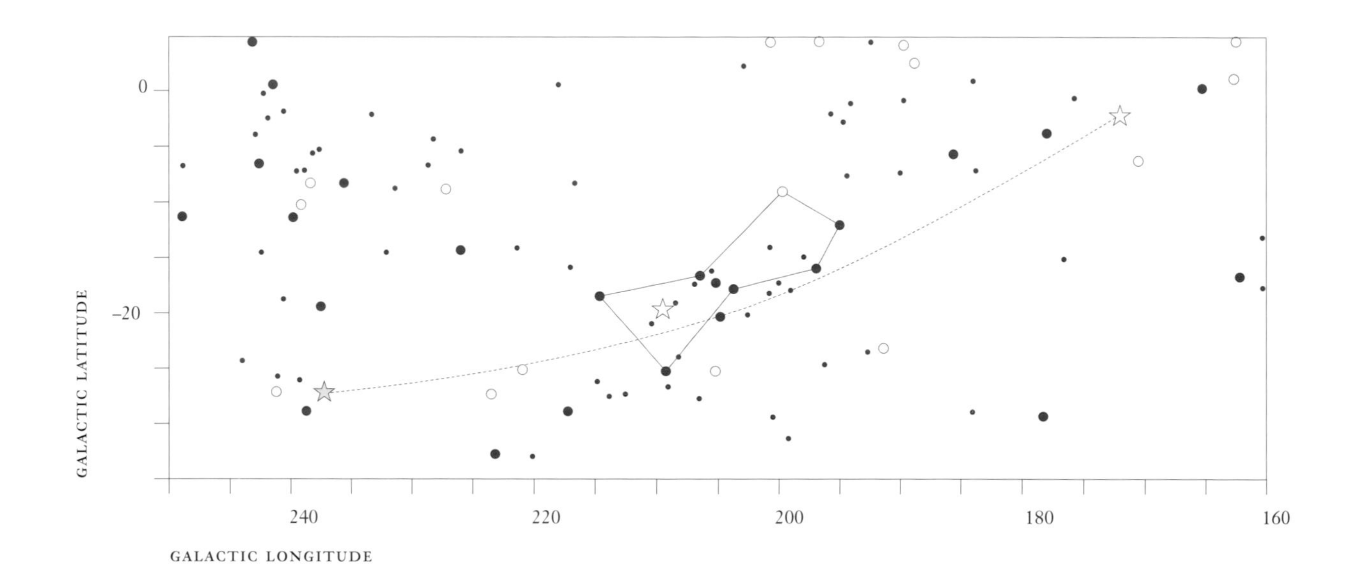

Two and a half million years ago, two double stars encountered each other near the Trapezium, exchanged stars, and ejected Mu Columbae (left) and AE Aurigae (right) in opposite directions, leaving Iota Orionis in the middle. *(Adapted from H. Hoogerwerf, J. de Bruijne, and P. de Zeeuw, from an article in* Astrophysical Journal Letters *544 [2000]: L133.)*

Residence: COLUMBA
Other name: HR 1996
Class: O9.5 DWARF
Visual magnitude: 5.17
Distance: 1300 LIGHT YEARS
Absolute visual magnitude: –2.88
Significance: THE FIRST KNOWN RUNAWAY STAR.

MU COLUMBAE

Mu Columbae is the first known runaway star. Or maybe it is the second known runaway star, as it was one of a pair. Over the last 100 or so years, astronomers have put an immense effort into knowing not just the distances of the stars, but also how they are moving. Measures of angular motion across the line of sight, called the proper motion, coupled with distance gives the velocity across the line of sight in kilometers per second. The motion along the line of sight is easily determined through the Doppler effect, which is found by measuring the minute shifts of absorptions in the stellar spectra. The combination of the two motions allows the true velocity vectors—speeds and directions relative to the Sun—to be known.

Oddballs clearly stand out from the rest. Among them, found 50 years ago, are Mu Columbae, south of Orion, and AE Aurigae, well north of the constellation. They are moving away from Orion in almost exactly opposite directions, separating from each other at over 200 kilometers per second. Moreover, AE Aurigae (a slight, irregular variable) is obliviously plowing through a recently encountered interstellar cloud (like **MEROPE** in the Pleiades), showing it to be an interloper. Together with 53 Arietis (northwest of Orion and added later), the three became the archetypal runaway stars, looking like uninvited guests that the Hunter kicked out of his home.

Mu Columbae would stand out all by itself though. A blue-white O9.5 star of nearly 15 solar masses 1300 light years away, it radiates nearly 21,000 solar luminosities from a 33,000-K surface. Not by coincidence, AE Aurigae and Mu Columbae have similar distances and identical spectral types, while 53 Arietis, a B1 dwarf, is closer at 760 light years and just a bit cooler and less massive.

Most surprising, there is nothing terribly unusual about such behavior. Roughly 20 percent of all O stars and nearly 10 percent of all B stars are runaways! What could be going on? How do you eject a massive star? For decades, astronomers argued over two scenarios. A large fraction of hot massive stars are in pairs. Were one star to explode off-center, the collapsed remnant would be given a kick so violent that it would leave the system at high velocity (*à la* **GEMINGA**). The companion, surprisingly undamaged by the event, would be kicked in the opposite direction, and off they go, never to rendezvous again. However, explosions are not needed. Massive stars are also found in dense star clusters. Random gravitational interactions among the members can accelerate stars, break up binary pairs, and launch stellar bullets into the surrounding cosmos.

When the motions of AE Aurigae and Mu Columbae are computed back in time along their Galactic orbits, they (along with Iota Orionis, a part of Orion's Sword) seem to have come out of the Trapezium, the quartet of stars that illuminates the Orion Nebula. The Trapezium is at the core of a great and dense stellar cluster, while Iota Orionis is an attractive multiple star. The main component of Iota Orionis, a third magnitude O9 star giant, is by itself a very close double whose similar members have unusually eccentric orbits about each other. Apparently two doubles within or near the Trapezium closely encountered each other 2.5 million years ago. Of the four, two were ejected, and two others fell into a weird mutual orbit.

Both mechanisms may work. Zeta Ophiuchi, an O9.5 dwarf, has been paired with a pulsar in Aquila over 45 degrees away. A supernova must have done that one. Stay alert, as you never know when a star might be fired in your direction.

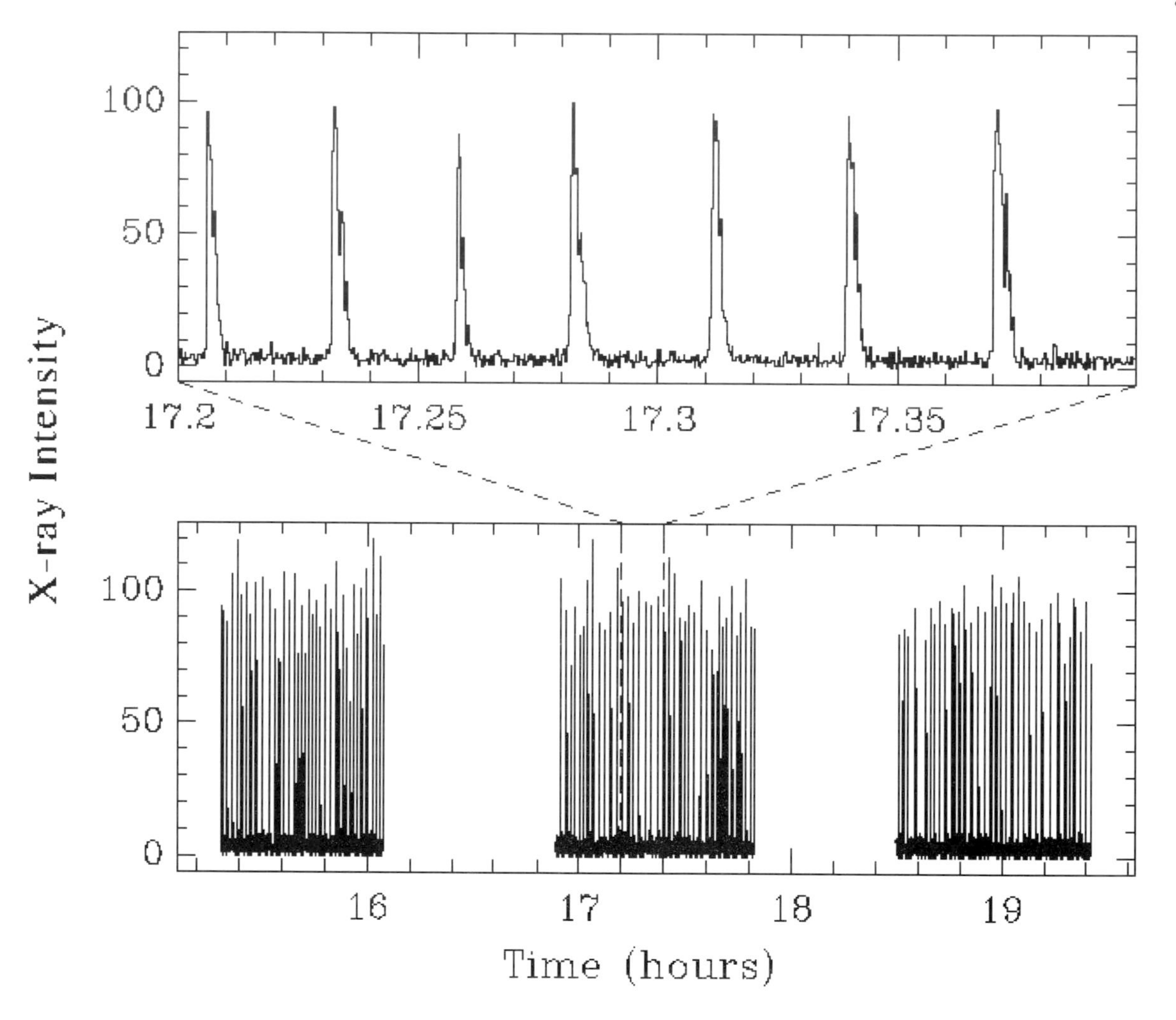

Staccato X-ray bursts emanate from the Rapid Burster during a three-hour-long observing period. *(BeppoSax [a joint Italian-Dutch program] X-ray observations, N. Masetti et al., from an article in* Astronomy and Astrophysics *363 [2000]: 188.)*

Residence:	SCORPIUS
Other name:	RAPID BURSTER
Class:	LOW-MASS X-RAY BINARY
Visual magnitude:	IRRELEVANT
Distance:	25,000 LIGHT YEARS
Absolute visual magnitude:	IRRELEVANT
Significance:	A WEIRD X-RAY BURSTER.

MXB 1730–335

MXB 1730–335 ("Massachusetts X-Ray Burster") is far away, near the center of the Galaxy. It is also a unique X-ray source. Buried within a globular cluster obscured by interstellar dust, it is quite hidden optically. The first X-ray source ever to be discovered was also in Scorpius. Both it and MXB 1730–335 are "ordinary" low-mass X-ray binaries, as opposed to the (surprise) high-mass X ray binaries, exemplified by **SS 433**. Each kind glows in X-rays when an ordinary star passes matter onto an orbiting neutron star (which was created earlier in a supernova explosion). The powerful gravity of the dense neutron star then compresses and heats the infalling gas. The neutron star in the HMXBs have class O and B companions that are greater than ten solar masses, while the LMXBs companions are under one solar mass. Nothing is known to fall in between.

About half the LMXBs are bursters that typically flash a powerful X-ray burst, which rises within seconds and then dies away in seconds to minutes. The bursts repeat in intervals from fractions of an hour to days. Temperatures of the flashing stars can hit 30 million K as they release X-radiation at a rate 100,000 times the total solar luminosity.

Novae like **DQ HERCULIS** are created by matter that flows from a low-mass dwarf onto a white dwarf. When the new hydrogen is sufficiently compressed, it explosively fuses to helium via the carbon cycle. LMXB bursts, or "X-ray novae," are similar. Here, matter is tidally drawn from a low-mass star into an accretion disk surrounding the neutron star. From there, it rains onto the neutron star's surface. The huge compressing gravity of the neutron star causes the fresh hydrogen to fuse continuously to helium. When enough helium has accumulated, it explosively fuses to carbon and oxygen (the same process that fires giant stars), which expands and cools the compressed atmosphere of the neutron star. A few minutes to days later, it happens again, depending on how much matter is dumped through the accretion disk. An LMXB burster is a repeating helium bomb.

The Rapid Burster does all this very nicely. However, it also produces the romantically-named "Type II" bursts (logically rendering the helium explosions "Type I"), which are lower-energy staccato machine-gun-like blasts that can last anywhere from a couple of seconds to a few minutes. The interval between them can be as little as seven seconds. Then the bizarre behavior goes away for a time, for weeks, only to pick up again. Only one other Type II source is known.

The origin of these strange bursts seems to lie within the accretion disk that is feeding mass to the neutron star. In the case of the "RB" (give it a nickname and it becomes friendlier, if not better understood), the accretion disk is probably very unstable. Filling at a regular pace from the ordinary dwarf, it suddenly and violently dumps its load, repeatedly doing so until it settles down for a time. Such weird stars have great importance. Though at first the oddballs of a stellar crowd may seem so rare that they can be ignored, they commonly give us physical clues whereby the whole set can be better understood.

The neutron star within an LMXB was probably once a pulsar similar to that found in the **CRAB NEBULA**, an old one that radiated its energy and is now spinning comparatively slowly. The accreting mass falls upon the neutron star at an angle, giving it repeated glancing blows and speeding it up. Eventually, as it consumes its partner, the neutron star will appear as a wildly revolving millisecond pulsar like the **BLACK WIDOW**, thus giving us yet another connection through which we can see the whole cycle of stellar evolution.

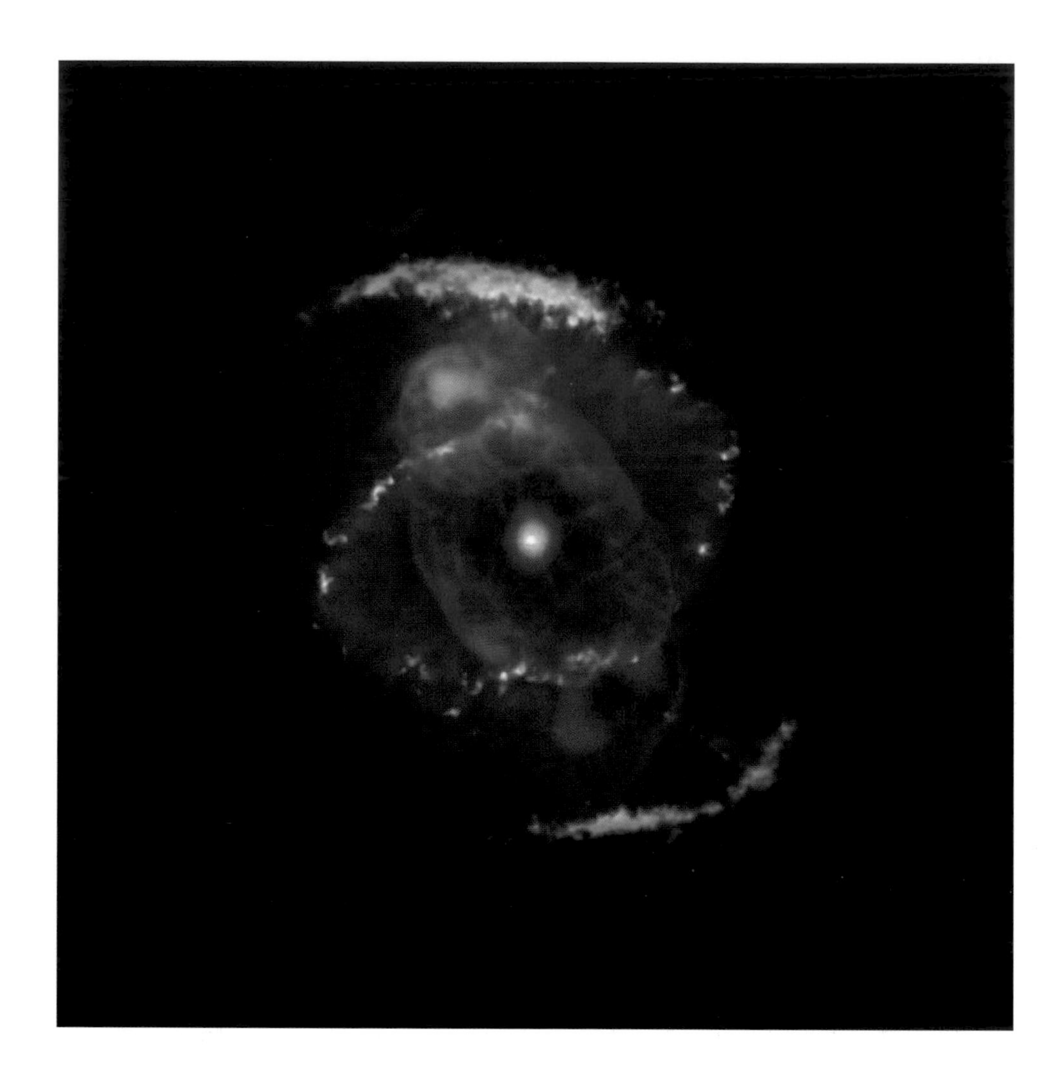

NGC 6543, the "Cat's Eye," one of the most complex and beautiful of planetary nebulae, was the first to be imaged by the Hubble Space Telescope. X-ray emission (colored blue) from superheated gas fills in the optically visible whorls. *(Hubble Space Telescope, J. Harrington and K. Borkowski [University of Maryland], STScI, and NASA; X-ray: Chandra X-Ray Observatory, Y.-H. Chu [University of Illinois] et al., and NASA.)*

Residence:	DRACO
Other name:	CAT'S EYE
Class:	PLANETARY NEBULA
Visual magnitude:	11.1 (STAR)
Distance:	3250 LIGHT YEARS
Absolute visual magnitude:	1.1
Significance:	A COMPLEX PLANETARY NEBULA WITH A LOT OF FIRSTS.

NGC 6543

In 1785, four years after discovering Uranus, William Herschel announced a small glowing body in Aquarius that, based on its disk-like appearance, he called a planetary nebula. It was followed by many more, including one in Draco within which Herschel discovered a star that completed the definition of the class: planetary nebulae are bright clouds that surround single central stars. The most famed planetary nebula is the Ring Nebula in Lyra, but of all of them, the Draco Nebula (which 100 years later became NGC 6543) is probably bemedalled with the most important "firsts." In 1864, William Huggins used his spectroscope to find bright colorful emissions that proved planetary nebulae to be gaseous. It was also the first of its breed to be observed with the Hubble Space Telescope, its awesome structure giving it the popular name "Cat's Eye Nebula."

Planetary nebulae are the ejecta of giant stars. A second-ascent giant like **MIRA** or **IRC +10 216** eventually loses its entire outer envelope through its powerful wind, some of which condenses to dust and temporarily hides the star within. As the star loses mass and begins to expose its nuclear-burning core, it heats. For a short while, the inner starlight is reflected from the expanding cloud's dust grains, and we see something like the **EGG NEBULA**. Finally, the heating star tops 26,000 K, at which point it releases ultraviolet radiation energetic enough to ionize the surrounding gas, stripping electrons from atoms. The rejoining of the electrons with their nuclei radiates energy in the visible spectrum, and lo: a planetary nebula is created. Eventually, the nebula will dissipate into space, leaving a white dwarf like **SIRIUS B** behind. While separate, the two are still one entity, a core whose outer fleeing envelope is just very large and transparent.

Though lasting only a few tens of thousands of years, planetary nebulae abound as a result of their brightness and easy visibility. The Cat's Eye, an uncertain 3250 light years away, is about a third of a light year across and is expanding away from its star at about 20 kilometers per second, implying an age of only 1500 years or so since ejection. The nebula, heated to a temperature of 8000 K, has a mass of about one tenth solar. Even at 47,000 K, the core of NGC 6543 is on the low side. Still heating, it will top 100,000 K before finally cooling to become a white dwarf. Though at 11th magnitude seemingly faint, the star emits so much ultraviolet that it is really very luminous, radiating 3600 times more energy than the Sun. Now cut down by mass loss from 1 solar mass to around 0.6 solar, it is somewhat deficient in hydrogen and may be on its way to becoming a helium white dwarf like **HZ 21**.

The Cat's Eye, like all other planetaries, is structured when a hot wind from the nascent central star slams into the older wind lost during its giant lifetime, and like a violent snowplow, shovels it forward. The star still blows a wind at over 2000 kilometers per second at a rate approaching a ten-millionth of a solar mass per year. Surrounding the inner nebula is the faintly glowing solar-mass old wind from the once-giant, an outer halo of equal complexity 1.5 light years in diameter. The original star must have had a mass half again as great as the Sun.

Unlike more extreme nebulae such as **NGC 7027**, NGC 6543 has solar chemical abundances. Convection within the evolving star was not effective enough to have brought freshly-made carbon, nitrogen, and the like up from below. Nevertheless, the nebula will add its load of gas and dust to the clouds of interstellar space, from which new generations will be born, firmly joining stellar death with stellar life.

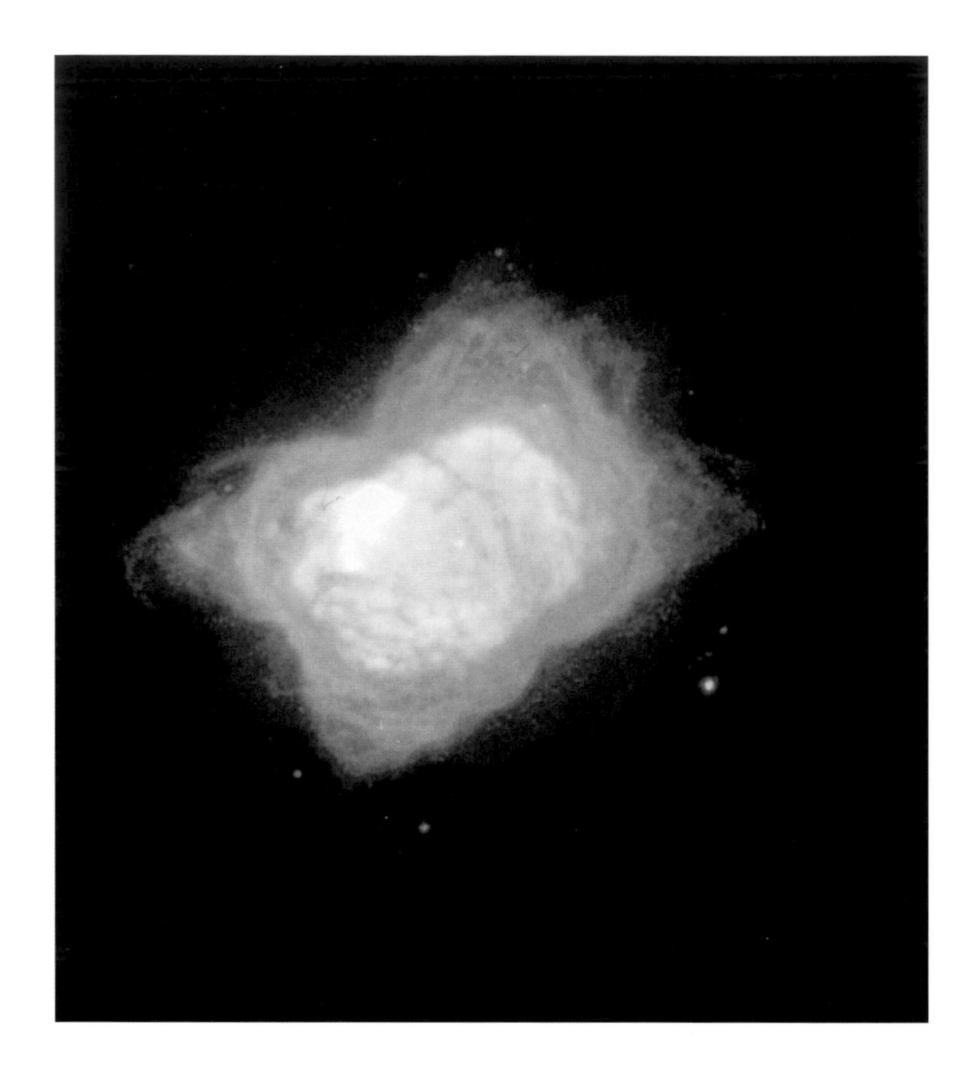

The planetary nebula NGC 7027, filled with dark lanes of absorbing dust and surrounded by molecular hydrogen, has one of the hottest central stars known. *(Hubble Space Telescope, H. Bond [STScI] and R. Ciardullo [Pennsylvania State University], A. Hajian [Naval Observatory], Y. Terzian [Cornell University], M. Perinotto [University of Florence, Italy], P. Patriarchi [Osservatorio Astrofisico di Arcetri, Italy], Nicmos Team, STScI, and NASA.)*

Residence:	CYGNUS
Other name:	PK 084–03 1
Class:	PLANETARY NEBULA
Visual magnitude:	16.3
Distance:	2900 LIGHT YEARS
Absolute visual magnitude:	6.6
Significance:	AN INTENSELY OBSERVED PLANETARY NEBULA WITH ONE OF THE HOTTEST KNOWN CENTRAL STARS.

NGC 7027

NGC 7027 has no actual popular name. This extraordinary and bright planetary nebula is so well-known that "NGC 7027" *is* a popular name. Everyone in the planetary nebula research field knows it, even if he or she has not done work on it. The nebula is the subject of some 50 research papers per year.

What makes an object so far removed from observatory open houses so professionally popular? Though its structure is hard to discern in a small telescope, NGC 7027 is among the visually brightest of the sky's planetary nebulae, and its spectrum is spectacularly rich with emissions from a great variety of atoms and ions. Observed in optical radiation, it has a weird shape far removed from the usual elliptical structures presented by most planetaries like **NGC 6543**. Much of the nebula is obscured by dust. Only recently was the central star, hidden by both dust and bright nebulosity, finally found.

However, NGC 7027 is no longer a mystery. Advanced optical, infrared, and radio imaging reveal one of heaven's more wonderful stellar bodies. The infrared punches through the light-absorbing dust to show a complex elliptical ring one tenth of a light year across, which is still axed by thick dusty lanes. Observing at wavelengths that avoid the powerful spectral emissions exposes a 16th magnitude star. Radio observations reveal a huge surrounding structure a full light year across that is filled with molecules. Infrared radiation displays a remarkable interface between the two domains.

NGC 7027 is a very young, carbon-rich planetary nebula. A mere 700 years ago, it was ejected from a **MIRA**-like, second-ascent variable star and still shows some of the dusty-molecular characteristics of its predecessor. Visually five times less luminous that the Sun, the central star seems so feeble as to be worth ignoring. It fools. With a spectacular temperature of 198,000 K, it radiates almost all its energy in the hard, ionizing ultraviolet and the X-ray regions of the spectrum. The nebula converts so much of that energy into optical radiation that the visually faint star nearly disappears. When the ultraviolet is included, we find the star to radiate at a rate 7700 times the solar value.

The vast molecular envelope is dominated by cold molecular hydrogen, which is protected from the destructive radiation of the ultra-hot central star, that is, the nebula absorbs the radiation before the surroundings can be penetrated and ionized. Expanding at about 20 kilometers per second, the bright inner structure is that portion of the cloud that has been shoveled by a fast wind from the central star and then ionized. While the visible nebula contains perhaps one tenth of a solar mass, the outer part holds close to three solar masses. Between the bright inner region and the molecules lies a complicated four-lobed transition zone. Here, the molecules are being broken up as the bright part eats its way into the dark envelope. In three dimensions, this structure looks like a double cone of the kind seen in the dust of the **EGG NEBULA**.

Theory shows that the more massive stars of planetary nebulae heat much faster than those of lower mass, which explains why NGC 7027's star can be so hot so shortly after nebular ejection. Its current estimated mass of around 0.7 times that of the Sun, plus all that it has lost, suggests an initial mass of 3.5 times solar, making NGC 7027 the daughter of a mid-class B dwarf. The star will probably become a helium-rich white dwarf like **HZ 21**, while the planetary nebula will expand to invisibility. As it dissipates, it will deliver another load of carbon to interstellar space, matter that will someday find its way into new stars and planets.

Nova Cygni 1975 erupted from a star that was previously so faint as to be nearly undetectable on this photograph. The seemingly extended image of the star after eruption is an artifact. *(© University of California Regents/Lick Observatory.)*

Residence:	CYGNUS
Other name:	V 1500 CYGNI
Class:	NOVA
Visual magnitude:	1.85 (MAXIMUM)
Distance:	5000 LIGHT YEARS
Absolute visual magnitude:	–10.7 (MAXIMUM)
Significance:	FAST, BRILLIANT MAGNETIC NOVA.

NOVA CYGNI 1975

In the summer of 1975, the Swan sprouted a second tail as Nova Cygni 1975 erupted into the night sky. Reaching nearly first magnitude, it rivaled **DENEB** and then quickly faded. Novae are common, with several discovered every year. Those bright enough to bring people out to look, such as Nova Persei 1901, Nova Aquilae 1918, and Nova Herculis 1934, occur about 20 years apart. After the eruption, the nova acquires a variable star name, Nova Herculis becoming **DQ HERCULIS**, Nova Cygni transmuting to V 1500 Cygni.

A nova is the product of a close double star where a white dwarf is married to a low-mass ordinary dwarf. The white dwarf so tidally distorts its companion that it can draw matter onto itself. When sufficient fresh hydrogen accumulates onto the white dwarf's surface, it explodes in a thermonuclear runaway. Within a couple of days, the binary can become as bright as the most luminous stars in the Galaxy. At first fading by a magnitude every ten days or so, the white dwarf declines slowly, taking centuries to cool and return to normal. Shortly after the outburst, the ejecta of the explosion become visible as an expanding nebula. Since no two couples will be exactly alike, neither will be their novae.

Beginning with an original magnitude fainter than 21, Nova Cygni rose from obscurity faster than any nova on record. "Fast" novae are brighter than "slow" ones, and Nova Cyg was no exception. It may have reached absolute visual magnitude –10.7 at its peak and a total maximum luminosity of 2.5 million times solar. It also faded faster than any other, almost one magnitude a day. Yet even now, the star is easily accessible. The white dwarf's temperature hovers around 100,000 K, and the luminosity, several times that of the Sun, is far in excess of an ordinary white dwarf.

More importantly, Nova Cygni was a "magnetic nova." Like **EG 129**, the accreting white dwarf is highly magnetized, although the field strength is unknown. Novae are extreme examples of cataclysmic binaries, of which there are several quieter kinds. All, like **SS CYGNI**, are close doubles that contain white dwarfs which disturb and accrete from their companions. In most cases, the accreting mass flows first into a disk around the white dwarf and then down onto its surface. Instabilities in the disk make the double star flicker and jump in brightness. In the AM Herculis subset, the white dwarfs have magnetic fields 10 million or more times the strength of the Sun's. The fields are so powerful that they prevent the accretion disks from forming and instead funnel the incoming gas directly onto the stars' magnetic poles. Also called "polars," they radiate X-rays and flip back and forth between high- and low-radiating states as a result of variations in the accretion rates. The magnetic fields are so strong that the white dwarf rotation periods lock onto the orbital periods. (DQ Herculis is an intermediate case in which the magnetic fields are not strong enough for synchronization and disk suppression.)

Nova Cygni 1975 was an AM Herculis star, and when it returns to normal, it will be one again. The explosion was so violent that it desynchronized the rotation from the orbital revolution. The white dwarf is now spinning every 3.29 hours, while orbiting with its low-mass companion every 3.36 hours. The two are gradually re-establishing their former intimate relationship. When the system quiets down and more hydrogen has been accumulated by the white dwarf, perhaps in hundreds of thousands of years, the two will have another spat, and the white dwarf will once again blow its top.

OH 231.8 +4.2 (the Calabash Nebula) is a double-lobed bipolar flow one light year long that emanates from a hidden MIRA variable buried in the "neck." It is illuminated by reflection from dust grains. The small end is pointed at us, the larger one away, as gas speeds outward at up to 300 kilometers per second. *(Hubble Space Telescope, ESA and V. Bujarrabal [Observatorio Astronómico Nacional, Spain], STScI, and NASA.)*

Residence:	PUPPIS
Other name:	CALABASH NEBULA; QX PUPPIS
Class:	M9 GIANT AND OH/IR STAR.
Visual magnitude:	STAR: OPTICALLY INVISIBLE
Distance:	4200 LIGHT YEARS
Absolute visual magnitude:	–5
Significance:	AN OXYGEN-RICH "MIRA" IN TRANSITION.

OH 231.8 +4.2

Pet names make astronomical bodies friendly and are a wonderful aid to memory. In the constellation Puppis lies a double-lobed cloud that looks like a gourd: hence OH/IR 231.8 +42 (the numbers representing Galactic coordinates) became the "Calabash Nebula." Such "OH/IR" sources abound. "IR" in the name tells us that the Calabash is a strong infrared radiator, while "OH" means that hydroxyl maser emission is present. Hydroxyl is a radical, an incomplete molecule (add another hydrogen atom and you get water) that cannot exist under terrestrial conditions. The density in circumstellar clouds is so low, however, that the stuff can be permanent. The maser is a microwave laser in which atoms and molecules are forced to radiate intense coherent radiation, in which all the waves travel in lockstep, producing a powerful beam. In the OH/IR sources, the maser energy is derived from the infrared radiation of a central **MIRA**-like variable.

Mira variables, all of which are second-ascent giant stars, come in three flavors: oxygen-rich M stars, carbon-rich C stars (**IRC+10 216**), and those in which oxygen matches carbon (S stars like **CHI CYGNI**). Clouds of ejecta surrounding carbon stars are laced with carbon molecules, while those surrounding the oxygen-rich M stars contain the OH and other oxygen molecules. Inside the Calabash, buried so deeply it is detected only in the infrared, there beats a cool class M9 "Mira" (QX Puppis) with a long, 700-day period.

The two lobes are huge bipolar flows that, in some places, reach speeds of 300 kilometers per second and are mostly illuminated by the reflection of internal starlight by dust grains. A small amount of atomic emission from the gas reveals shock waves that are generated as fast winds slam into slower-moving gas. The outer edges of the lobes are Herbig-Haro objects similar to those created by jets from younger bodies like **L 1551 IRS 5**. The lobes of the Calabash must be focussed by some sort of disk structure, as are the jets in young stellar objects. Stellar death, we see, replicates stellar birth!

The OH radiation in the nebula responds to the Mira variations, but with a delay of the light-travel-time from the star to the OH-radiating region. The speed of light gives the region's size, and comparison with the angular size yields distance (reflection from embedded dust grains works too). The Calabash is 4200 light years away and roughly 1 light year long. The luminosity of the embedded star is around 10,000 times that of the Sun, consistent with the mass required by such a long pulsation period. QX Puppis probably began its life as a three-solar-mass main-sequence B star and is still in the process of evaporating its outer envelope. The lobes, the ejecta from the massive star, contain somewhere around a full solar mass, all lost from buried QX Puppis. Based on their size, the massive outflow should have begun around 1000 years ago.

OH 231.8 +4.2 is unique. The flow of late stellar evolution is from "Miras" to "protoplanetary" (nascent) nebulae like the **EGG NEBULA** (which usually have warm class F and A stars buried within them), through full-blown planetary nebulae (**NGC 6543** and **NGC 7027**), to white dwarfs (**SIRIUS B**). The Calabash has the same kind of biconical structure that is found in both the Egg and in NGC 7027, helping to link them all together. Here may be the crossover, the connection of the second-ascent giant star to the protoplanetary stage. QX Puppis is not quite a nascent planetary nebula, nor is it really quite a Mira variable, but one teetering right on the edge of its final path to death. We can see both where it came from and where it is destined to go.

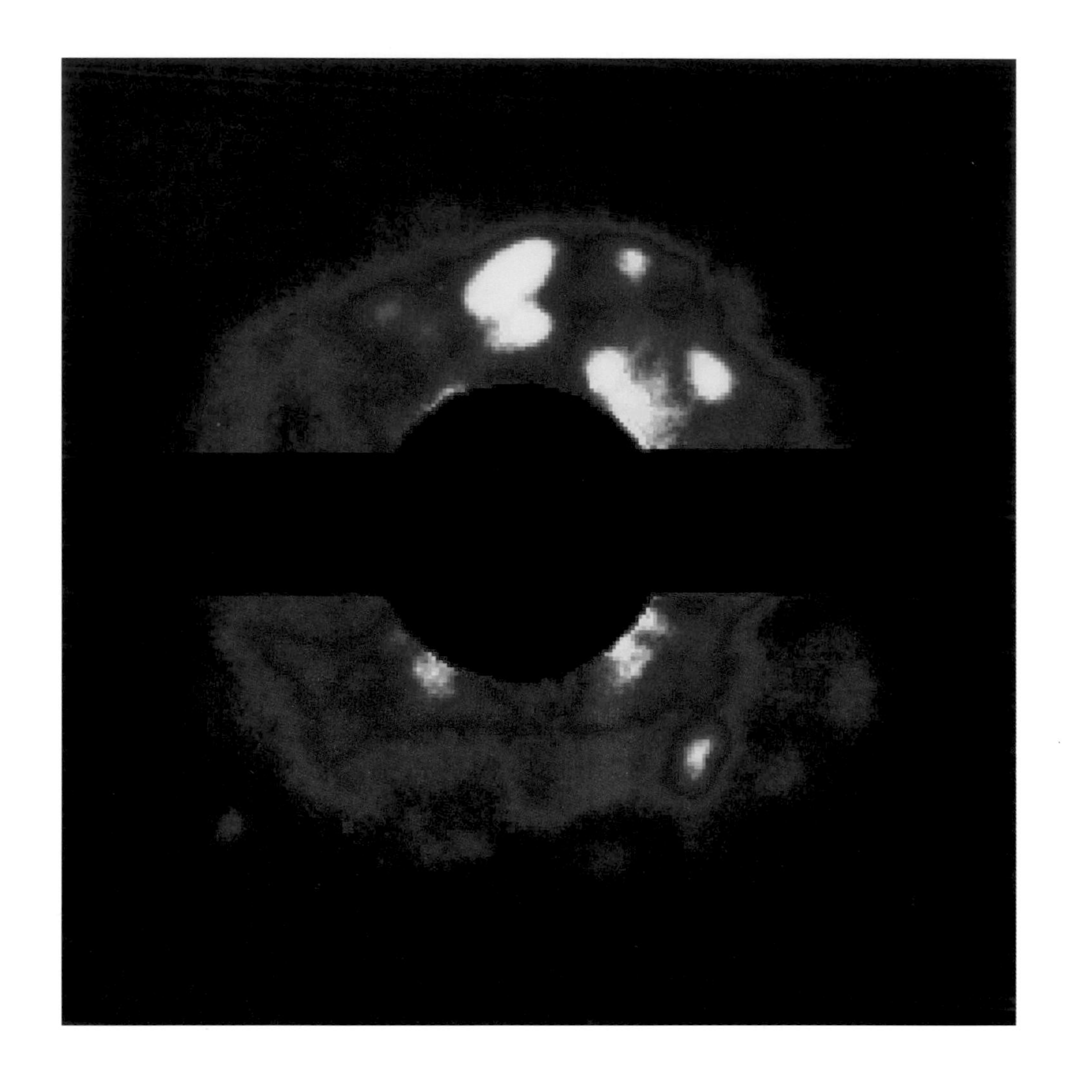

A cloud of gas ejected by P Cygni surrounds the star, whose bright light is blocked by a disk within the telescope. *(Hubble Space Telescope, A. Nota, STScI, and NASA.)*

Residence:	CYGNUS
Other name:	NOVA CYGNI 1600
Class:	B2 HYPERGIANT
Visual magnitude:	4.8
Distance:	6000 LIGHT YEARS
Absolute visual magnitude:	–8
Significance:	THE EPONYMOUS WINDY STAR AND RARE LUMINOUS BLUE VARIABLE.

P CYGNI

When Johannes Bayer ran out of Greek letters while naming the stars, he moved on to lower-case Roman letters and then to upper case, finally giving up at Q. Though the Roman letters are rarely used, they do pop up from time to time: a cluster in Perseus is still called "h Persei," and a fairly bright star in Scorpius is tagged with "G." The best-known of these lies in Cygnus, Bayer's "P," an amazing star that has lent its name to a special spectral phenomenon.

At fifth magnitude, P Cygni does not make much of an impact on its constellation, a celestial Swan set beautifully in the Milky Way. In reality, however, it is among Cygnus's most luminous stars, indeed among the most luminous in the Galaxy. From estimates of distance between 6000 and 7000 light years and with a temperature of 18,000 to 19,000 K, it radiates between 500,000 and 900,000 times the solar energy. While not up to ETA CARINAE's vast standards, P Cygni certainly is in the same league.

P Cyg is famed on two other counts. In 1600, it flared to third magnitude and was thought to be a nova. Instead, it ended up behaving something like a miniature Eta Carinae. The eruption went on for six years, and then the star faded below naked-eye visibility, only to rise again for several years in 1654. Since settling in at fifth magnitude a century later, it has slowly increased its brightness (with many superimposed variations) by about 15 percent per century. The increase was not intrinsic, but was the result of cooling by 6 percent per century, which transfers progressively more of the star's ultraviolet light into the visible.

Of equal interest is P Cygni's remarkable spectrum. The dramatically different spectral features are composed of emissions flanked to the short-wave side by deep absorptions. In a normal star, the dark gaps in the spectrum are produced by the different chemical elements and ions in the star's semi-opaque atmosphere absorbing outflowing radiation at specific wavelengths. P Cygni, however, blows a powerful wind that partially hides the star. Gas flows out in all directions, some across the line of sight, some directly at us. The part of the wind not seen against the star radiates the bright emissions. The portion of the wind coming at us is seen against the star and creates absorptions. The line-of-sight velocity of this directly on-coming gas equals the full wind speed, so the absorptions are Doppler-shifted to shorter wavelengths. "P Cygni lines" are seen in the spectra of a wide variety of stars, from new T TAURI stars to the dying central stars of planetary nebulae. They are enormously useful in allowing a measure of a star's mass-loss rate and wind speed.

P Cygni is losing mass at a high rate of three hundred thousandths of a solar mass per year (or even greater), 300 million times the solar rate, at a low speed of under 300 kilometers per second. Hence the wind is dense, and hence the P Cygni lines; mass loss can be episodically eruptive, however, hence the "false nova" of AD1600. The star is surrounded by a faint nebula that has been created over the past 900 years by the current eruptive mass loss and by faint shells that tell of eruptions from 2400 and 20,000 years ago.

The spectral class, the eruptions, and the luminosity place P Cygni among the hypergiant "luminous blue variables" that are ruled by Eta Carinae. With a mass between 50 and 60 solar, the star falls at the lower end of the lot. Stars from around 60 solar masses and up lose so much of themselves that they cannot evolve to become red supergiants, and instead stay blue supergiants until they explode. Though P Cygni could go either way, its eruptive and explosive future seems assured.

69

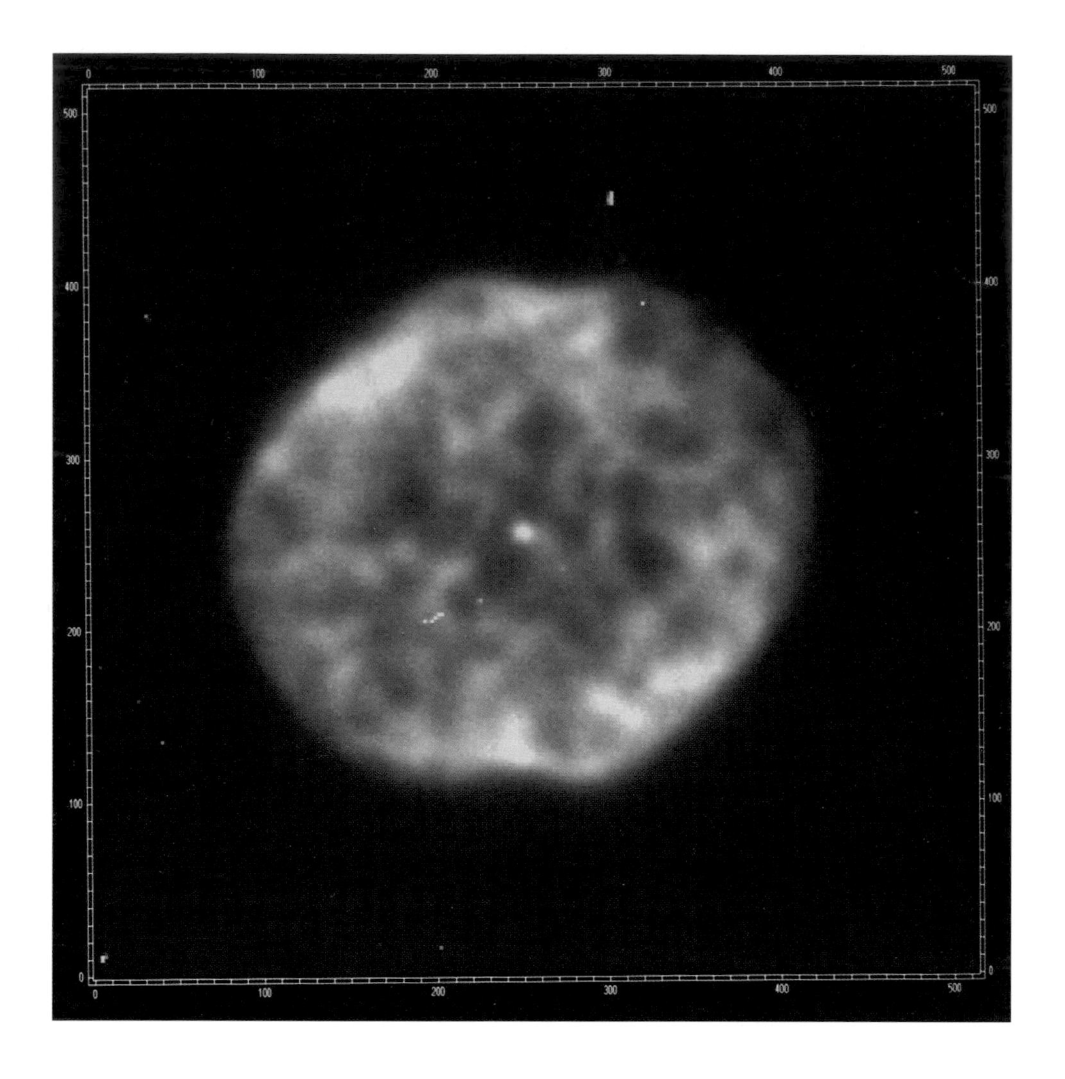

The central star of the planetary nebula NGC 1501, varying by up to a tenth of a magnitude with a typical period of 25 minutes, is a chatterbox star in the mold of PG 1159–035. *(NOAO/AURA/NSF, courtesy of Y.-H. Chu and G. Jacoby.)*

Residence:	VIRGO
Other name:	GW VIRGINIS
Class:	PG 1159 WHITE DWARF
Visual magnitude:	14.87
Distance:	600 LIGHT YEARS
Absolute visual magnitude:	6
Significance:	THE PROTOTYPE OF HOT, VARYING, PRE-WHITE DWARFS.

PG 1159-035

With some exceptions, we do not see stars evolve from birth to death; the time scales are too long. Instead, we take inventory of the different kinds of stars and mount them, like pearls, on a string of theory. Long lifetimes for any state of evolution yield vast numbers of stars, while short lifetimes yield few. Yet it is often the short transition stages that tell us much about how stars develop. As a result, astronomers will make great effort to study a group that has but a handful of members.

PG 1159–035 is the hot (130,000 K) prototype of about known 30 pre-white dwarf stars. Its luminosity of around 100 solar places it and its cohort in the realm of the central stars of planetary nebulae. And yet PG 1159 has no nebula. It was, however, found to be variable, subtly chattering through two hundredths of a magnitude with periods of both 7.7 and 9.0 minutes, and was given the variable star name GW Virginis.

To study the variations more completely, PG 1159 was a target of the unique "whole-Earth telescope," when nine observatories around the world continuously observed it for over 260 hours. The result was a measure of the strengths of over 100 different periods of pulsation! Such pulsations, which depend on stellar parameters, gave a mass of 0.59 solar (typical for a white dwarf) and a rotation period of 1.4 days.

The stars that belong to PG 1159 are characterized by absorptions of ionized helium and highly ionized carbon and oxygen. About half vary in the GW Virginis style, and about half (the two halves overlapping) are central stars of planetary nebulae, firmly linking the two kinds of stars together. Launched by second-ascent **MIRA** variables, the central stars of planetary nebulae heat at a constant high luminosity (over 1000 solar) as their outer skins are whittled away and the ancient nuclear-burning cores become more exposed. As they hit 100,000 to 200,000 K (the specific temperature depends on stellar mass), they begin to shut down residual nuclear fusion, while cooling and dimming to become white dwarfs. The PG 1159s are near this turning point. We have no idea, however, why some PG 1159s have no surrounding nebulae.

All classes of pulsating stars define domains bounded by limits of temperature and luminosity. The PG 1159 pre-white dwarfs fall between 80,000 and 170,000 K (though numerous stars that are not PG 1159s also fall in this realm). As the PG 1159s cool and dim to become real white dwarfs, they pass through another instability domain between 22,000 and 28,000 K. Here, we find the helium-rich version of the hydrogen-rich (DA) **ZZ CETI** white dwarfs, which occupy a belt between 11,300 and 12,500 K. Classical Cepheids (like **DELTA CEPHEI**) are forced to pulsate by the valving of radiation in a deep layer in which helium is becoming ionized. The ZZ Ceti stars pulsate as a result of similar valving by hydrogen, the DB pulsators by helium, and the PG 1159s from the same effect produced by oxygen and carbon. Theory thus links diverse sets of stars in different states of evolution, unity derived from study of just a few.

At the end of the handle of the Little Dipper in Ursa Minor, Polaris, the North Star (near top center), lies less than 1 degree from the sky's north rotation pole. *(J. B. Kaler.)*

Residence:	URSA MINOR
Other name:	ALPHA URSAE MINORIS
Class:	F5 BRIGHT SUPERGIANT
Visual magnitude:	1.98
Distance:	425 LIGHT YEARS
Absolute visual magnitude:	–3.59
Significance:	THE POLE STAR, THE DEFINING SECOND MAGNITUDE STAR, AND THE SKY'S BRIGHTEST CEPHEID VARIABLE.

POLARIS

Polaris, the North Star, the Star of the Sailor, guides our way above all others. At the end of the handle of the Little Dipper, Polaris anchors the tail of Ursa Minor, the Smaller Bear. Shining close to the North Celestial Pole, it guides us north and establishes the compass points. Apparently unmoving, it hangs to the north at an angle above the horizon equal to latitude. Walk toward it, and it climbs. When it stands over your head, you are at the cold North Pole of the world. From here radiate all the north-south lines of longitude that help give position on the Earth.

Nothing, however, is stationary. Actually half a degree off the true Pole, Polaris, like all other stars, traces a daily circle as the Earth rotates beneath it. Its polar residence is also temporary. As the Earth's axis wobbles over its 26,000-year precessional period, the true pole will draw closer to Polaris until about the year 2100 and then will begin to move away. In 13,000 years, Polaris will be 47 degrees away from the celestial pole (double the Earth's axial tilt), and the star will pass nearly overhead of Paris or New York. Our North Star is but one of a succession of "Polarises" that include **THUBAN**, which marked the Pole for the ancient Egyptians, and **VEGA**, the first magnitude luminary that will guide us when New Yorkers see our Polaris near their zenith.

Hardly the sky's brightest star, Polaris almost exactly defines second magnitude. Its rank of 41st is no fault of its own, but of its 450-light-year distance. In truth, the star is a magnificent six-solar-mass supergiant that shines 2200 times more brightly than our Sun. Its yellow-white light pours from a 7000-K surface 32 times the solar size. While Polaris circles the Pole, it is itself circled by two small companions—one of ninth magnitude a bit warmer than the Sun; another is tucked up near it with a 30-year period "visible" only through the Doppler effect.

Far outshining the prototype **DELTA CEPHEI**, Polaris is the not only the sky's brightest Cepheid variable, it is one with a difference. Cepheids are all brilliant yellow (class F or G) unstable pulsating supergiants that change their brightnesses and radii over intervals usually measured in days. Most vary by an obvious magnitude or more. Sneaky Polaris, however, changes by a mere 0.03 magnitudes (about 3 percent) over a 3.97-day period, the effect entirely invisible to the eye. Odder still, the variations themselves vary, having decreased from a tenth of 1 magnitude 100 years ago, something respectable Cepheids do not do.

For a time it was thought that the variations might actually stop, but the decline has clearly ceased. What is really happening, however, is just as interesting. As musical instruments have "fundamental" tones that give them their basic pitches, they also vibrate in higher-frequency overtones, which give them their timbres. Comparison with other Cepheids shows that Polaris is pulsating not with its natural fundamental period, but in its first overtone! Such overtone Cepheids tend to have lower amplitudes, though no one knows why this star's variations have decreased. We suspect that Polaris is now changing into its fundamental period of 5.7 days to become a more normal, higher-amplitude Cepheid. All we can do is wait, most likely for a time far longer than the 26,000 years it will take for it to become the Pole Star again.

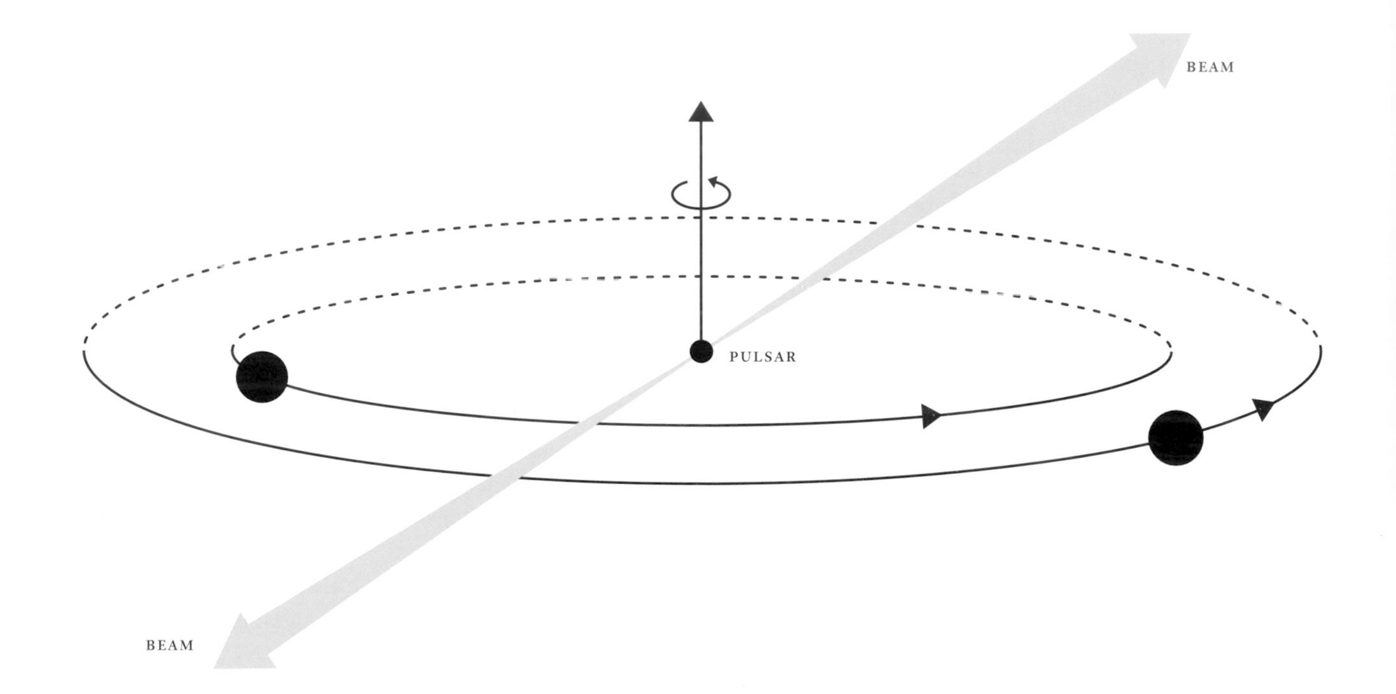

Apparently created from the debris of a destroyed star, two Earth-like planets orbit the millisecond pulsar PSR B1257+12 as the pulsar's twin beams swing wildly in space 161 times per second. Another planet orbits even closer, yet another one farther away. *(Adapted from J. B. Kaler [1994].* Astronomy! *New York: HarperCollins)*

Residence:	VIRGO
Other name:	NONE
Class:	PULSAR
Visual magnitude:	NONE
Distance:	1000 LIGHT YEARS
Absolute visual magnitude:	NONE
Significance:	FIRST PULSAR KNOWN WITH ORBITING PLANETS.

PSR B1257+12

Millisecond pulsars, old pulsars that have been spun up by matter accreting from binary companions, are the best natural clocks known. Spinning at speeds far greater than youngsters like the **CRAB NEBULA PULSAR**, most would be but a blur to the eye. The **BLACK WIDOW** rotates and blasts radio pulses toward us 640 times per second. A slower one, PSR B1257+12, turned the planetary world upside down.

The millisecond pulses arrive in a steady train with enormous regularity. If the pulsar is in mutual orbit with another body, it will be shifted back and forth along the line of sight. As it moves away from us, the pulses have progressively farther to go and thus arrive a bit late. As it approaches Earth, they arrive early. Like light of specific wavelengths, the pulses become "Doppler-shifted." The degree of the shift yields the velocity of the pulsar as it swings around its companion. That and the orbital period give information on the companion's mass and orbit. Millisecond pulsars are so accurate that minute masses can be detected.

PSR B1257+12 wobbles erratically. It has not one companion, but a pair. Better still, they are not stellar, but have masses not much greater than that of Earth. Pulsars can have planets! The outer one, of at least 2.8 Earth masses and 0.47 AU from its star (not much greater than Mercury is from the Sun), takes 98 days to orbit. The inner body (at least 3.4 Earth masses) takes 67 days at a distance of 0.36 AU. But are they real? Might they not be just artifacts of variations in the central pulsar? Two planets in a system must affect and perturb each other. The mutual movements change each others' paths in a predictable manner—just what is observed here. There really *are* two planets orbiting PSR B1257+12. Subtle variations in the timing later revealed a possible third, with at least 0.015 Earth masses at a distance of 0.19 AU. And we think there is even a fourth, even smaller one with a period of 3.5 years.

Of all things, a pulsar was the first star found to have an orbiting planetary system. How are they possible? Remember the Black Widow. She, the pulsar, is evaporating her mate. What happens to the debris? Some of it will fall onto the neutron star, but some will also go into orbit and will consolidate into planets. Given a source of raw materials, planets will apparently form almost anywhere, even around a pulsar, strong evidence that numerous planetary systems must orbit normal stars. But do not expect life. The high temperatures of neutron stars would make the planets much too hot. Even if they were not, a single blast from a rotating pulsar would be lethal.

While there have been other candidates, only two pulsar planet systems have been affirmed (the other one in the globular cluster Messier 4). Though rare, they tell us that we can look for planets most anywhere.

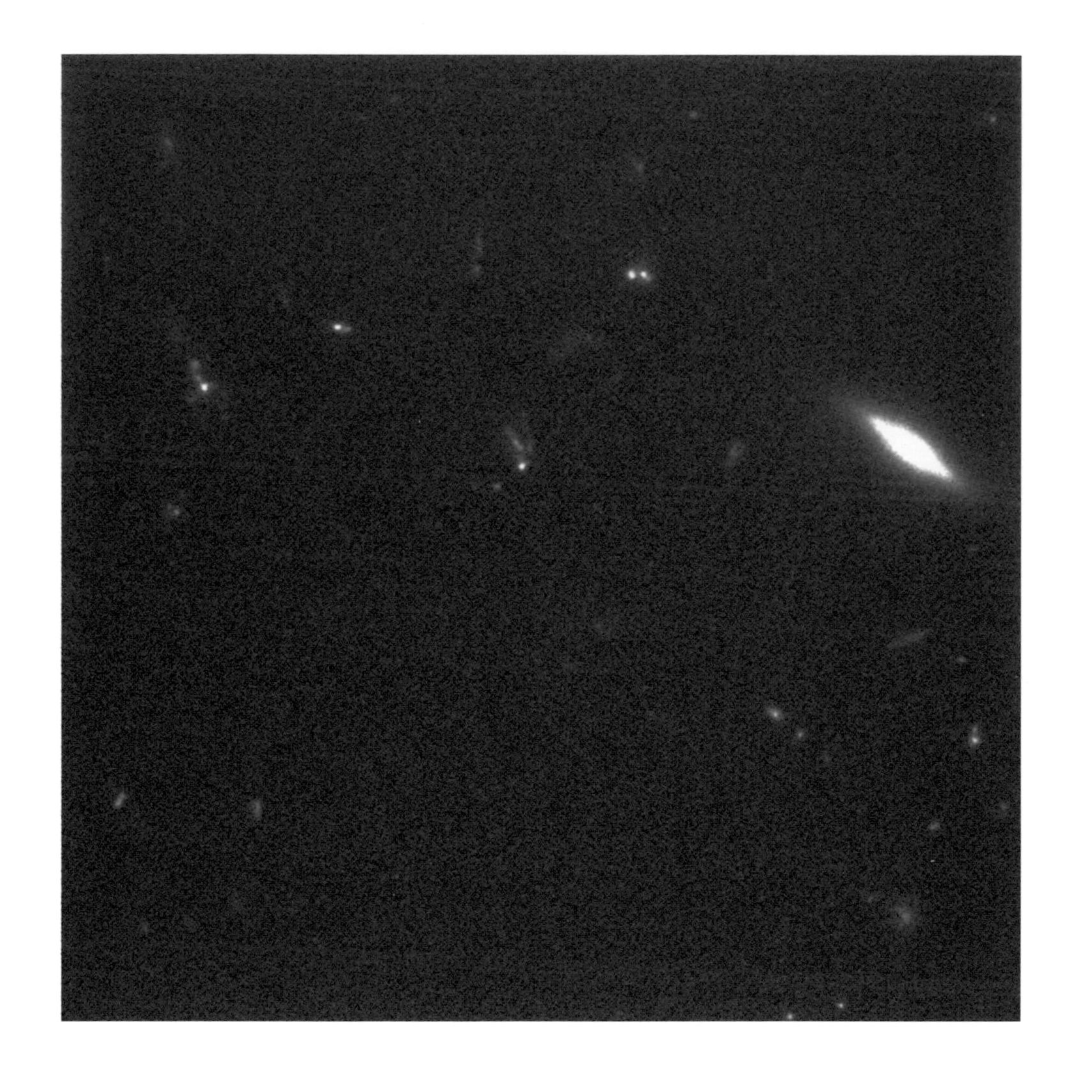

The optical glow from a powerful gamma-ray burst is seen in a distant galaxy (just above center) by the Hubble Space Telescope. Could it have been caused by the merger of twin neutron stars? *(A. Fruchter [STScI] and NASA.)*

Residence:	AQUILA
Other name:	HULSE-TAYLOR BINARY PULSAR
Class:	PULSAR + NEUTRON STAR
Visual magnitude:	> 22.5
Distance:	16,000 LIGHT YEARS
Absolute visual magnitude:	NOT KNOWN
Significance:	DOUBLE NEUTRON STAR THAT CONFIRMED RELATIVITY THEORY.

PSR B1913+16

Around 1870, the French astronomer Urbain Leverrier noted something peculiar about Mercury's orbit. The gravitational effects of the other planets should make Mercury's perihelion point (where it is closest to the Sun) move eastward by 527 seconds of arc per century, 43 seconds of arc slower than observed. No one knew why until 1914, when Einstein unveiled his theory of relativity. We think of time and space as two entities. Einstein locked them together in a unit called "space-time." Newton saw gravity as a force that extended over space to bring two bodies together. Relativistic gravity is caused by a distortion of spacetime that in a loose sense makes two bodies "roll together" and also solves Mercury's problem. The difference between Newton's and Einstein's theories increases along with the strength of the gravitational field. We could make a far better test of relativity if we could take our experiments to the powerful field of a neutron star.

PSR B1913+16 is a faint radio pulsar that bleeps rapidly (as a result of its spin) with a period of 53 milliseconds, just a bit slower than the **CRAB NEBULA**. It is, however, hard to keep up with, as the pulse period rapidly changes. The spinning pulsar is obviously orbiting something, its movement causing the pulse frequency to Doppler-shift back and forth with a period of 7.75 hours. (Mercury takes 88 days!) Analysis of the pulse arrival times gives the orbit's nature. The pulsar has a mass 1.44 times that of the Sun, the most precisely measured mass of any pulsar. The two bodies are separated by 2 million kilometers, just 1.4 times the solar diameter. The eccentric orbit brings them as close together as a solar radius. The position of closest approach rotates at an amazing 4 degrees per year, compared with Mercury's rate of 0.15 degrees per century. The double-star system accordingly provides a treasury of relativistic information.

Neutron stars are only 20 or so kilometers across, so even though the bodies are close together, each pulsar is effectively a point that gravitationally acts differently than would an extended normal star distorted by rotation or tides. Assuming that both bodies are points, the rotation rate of the orbit is exactly what relativity predicts. The companion's mass of 1.39 times solar is so similar to the pulsar's that, in addition to the lack of radiation from a more-ordinary star, it must be a neutron star as well. Its pulses might just miss the Earth, or it may not radiate at all.

Relativity also predicts that accelerating (here orbiting) bodies should produce "gravity waves" that dissipate energy and force the bodies closer together. While such gravity waves have never been directly observed, the orbit of the twin neutron stars is decaying exactly in accord with relativity theory, showing that this mysterious form of radiation does indeed exist. The observations also check a number of other relativistic parameters and rule out a variety of competing gravitational theories.

Two neutron stars take two supernovae. The originals should have been 16- to 18-solar-mass stars in fairly tight orbit. The more massive expanded first as a supergiant, lost mass to the other, and exploded as its core collapsed to become a neutron star. When the other expanded, it enveloped the new neutron star, and the two spiralled closer together. Then, the second supergiant blew up and formed another neutron star very close to the first, the violence putting them in eccentric orbit. Gravitational radiation will eventually cause the two neutron stars to merge, perhaps producing a burst of gamma rays that would be visible even from distant galaxies.

The symbiotic star R Aquarii, a MIRA variable coupled to a white dwarf, has ejected a nebula over half a light year across. (*© Anglo-Australian Observatory, photograph by David Malin.*)

Residence:	AQUARIUS
Other name:	HR 8992
Class:	M7 PECULIAR GIANT
Visual magnitude:	5.8 TO 11.5
Distance:	650 LIGHT YEARS
Absolute visual magnitude:	–0.7 TO +5.1
Significance:	A MYSTERIOUS ERUPTING SYMBIOTIC STAR.

R AQUARII

R Aquarii, the "first variable" in that constellation (the first discovered receiving the "R" designation, the second "S," etc.), is at first glance a common **MIRA**-type variable. It ranges from 6th to 12th magnitude and back over a 387-day period and has a spectral class that concomitantly changes from cool M5 to chilly M8.5. As a "Mira," it is also a second-ascent giant with a now-defunct carbon-oxygen core: a mass-losing star destined to become a planetary nebula like **NGC 6543**. R Aquarii's luminosity is difficult to assess, as the measured distance of 650 light years is accurate only to 50 percent. The star is close enough, however, for us to see it as a disk and to measure an angular diameter of 0.017 seconds of arc, which leads to a true diameter of 3.3 AU, slightly larger than the orbit of Mars. From that and the spectral class, the star is found to radiate around 5000 times as much energy as the Sun from a cool 2600-K surface.

R Aquarii, however, is no ordinary "Mira." It is also a "symbiotic star," one that shows two incompatible features: its spectrum contains M-star molecular absorptions coupled with bright emissions from highly excited ions and an unusual component of blue radiation. The temperature of the M giant is far too low to produce either, so a never-been-seen hot star has to be present. The picture from intense observational scrutiny that includes analysis of circumstellar matter shows a 1.75-solar-mass "Mira" in orbit about a mysterious 1-solar-mass white dwarf (or pre-white dwarf) with a temperature close to 100,000 K and a size perhaps ten times that of Earth. Eccentric paths take the pair from a maximum separation of 30 AU to as close as 5 AU, the latter distance being only three times the Mira star's radius. Maser (radio laser) emission from silicon monoxide shows that the "Mira" takes 5 to 18 years to rotate. Weirdly, detailed modeling of the system suggests an 18-year rotation period even for the small hot star, alluding to synchronous tidal locking. The hot star may also have a magnetic field hundreds of millions of times greater than Earth's.

The two stars must interact, and they do. The blue light occasionally brightens, and for a time—as it did in the 1920s and 30s—it can dominate the system when **P CYGNI**-type absorptions showed great mass outflows. R Aquarii is also surrounded by a unique nebula two minutes of arc across and aligned in east-west direction. A prominent jet screaming to the northeast contains matter flowing at over 200 kilometers per second. Another flow projects in the opposite direction. Motions within the flows suggest that the jet is about 100 years old.

Here is an extreme **ALGOL**-type system in which matter is being transferred from the big star to the small one, either through tidal distortion of the big fellow by the little one, or simply through the big star's wind. Unlike Algol, however, the stars are far enough apart that instead of landing directly on the hot star, the infalling matter first enters a flat disk and then rains downward. At the same time, some of it is expelled in the perpendicular direction. The disk around the white dwarf is apparently seen edge-on and is so thick that the little star is buried and hidden from us. The accretion flow rate is modulated by the eccentric orbit, and the disk is refueled every 44 years when the two stars approach their minimum distance apart. The varying distance thus causes periodic eruptions. The system seems capable of much greater, almost nova-like outbursts and has even been related to a "guest star" recorded in Korea in 1073, when R Aquarii perhaps produced its extended nebulosity. What, we wonder, will happen next?

R Coronae Borealis, just inside the curve of the Northern Crown, will unpredictably—and quite suddenly—disappear from view. *(J. B. Kaler.)*

Residence:	CORONA BOREALIS
Other name:	HR 5880
Class:	G0 SUPERGIANT
Visual magnitude:	4.9 TO 15
Distance:	4700 LIGHT YEARS
Absolute visual magnitude:	–5
Significance:	DISAPPEARING "SOOT STAR."

R CORONAE BOREALIS

A pretty semi-circle of stars, Corona Borealis (the Northern Crown) graces northern spring and summer skies. Within it lies one of the oddest and rarest of stars, one seen rather easily with the naked eye. At least most of the time. About once a year, or every several years, it disappears. Then you need a telescope, as the star will have sunk to magnitude 12 or even as faint as 15, which would make the star 10,000 times dimmer than normal.

Welcome to the bizarre world of R Coronae Borealis and its company of about 40 similar stars. They are mid-temperature, typically class G supergiants. None is close enough for decent parallax measure, the best limit on parallax placing R Cor Bor (as it is popularly known) at least 2500 light years away. The R Cor Bor stars that lie in the Magellanic Clouds—nearby galaxies whose distances we know—have absolute visual magnitudes of –5 (10,000 times the solar luminosity). If R Cor Bor were similar, it would have to be 4700 light years away to appear as faint as it does.

The star's variations are hopelessly erratic. It can go for years near maximum luminosity. Then, it will suddenly dive, sometimes taking a year to reach minimum, sometimes plunging in just a few weeks. At this point, it might just say, "Who, me?" and pop right back up as if nothing had happened. Alternatively, it might stay at a low level for years while making only brief excursions to resurface as a binocular star. Between fading events, the star pulsates slightly and slowly with a period of tens of days. The light curve (a graph of magnitude against time) is such that R Cor Bor has been called an "inverse nova," an exploder turned upside down.

R Cor Bor stars, however, have nothing whatsoever in common with novae. While stars like **NOVA CYGNI 1975** can explode as a result of surface nuclear fusion reactions, or brighten like **ETA CARINAE** as they erupt and eject matter, stars cannot suddenly fade. Something must be getting in the way. Since the catastrophic dimming is entirely irregular, it cannot be the result of an eclipse by an orbiting companion, as in the case of **EPSILON AURIGAE**. The only real explanation is the sudden expulsion of dust.

The key lies in the spectrum. The hydrogen absorptions are weak, revealing that the star is hydrogen-deficient and, therefore, helium-rich. Further analysis shows that the solar 2:1 oxygen-to-carbon ratio is reversed. R Cor Bor, a smoking celestial chimney, is expelling carbon-rich gas that turns to soot! It apparently does so in localized puffs: if one is in the line of sight, the star dims, if not, we notice nothing. Spectral evidence suggests that the soot forms close to the star, but no one really knows. Aeons of ejections have enmeshed the star in a warm, infrared-radiating cloud.

R Cor Bor may be a successor to **FG SAGITTAE**. FG Sagittae is the central star of a planetary nebula, one that was on its way to becoming a white dwarf. FG's striking changes in chemical composition indicate that it is fusing helium. As a result, the star is expanding and reverting into a helium-carbon-rich giant or even supergiant, an R Cor Bor star in the making. R Cor Bor's low mass of about 0.8 solar (found from pulsation theory) strongly supports this view, as only the stars that emerge from **MIRA** variables to become planetary nebula central stars can be this bright at such low mass. Long ago, R Cor Bor lost most of its hydrogen-rich atmosphere and cycled by-products of nuclear fusion (carbon and other elements) to its surface. It will eventually make the long slide back to white-dwarfhood, its last gasp at glory at best ephemeral.

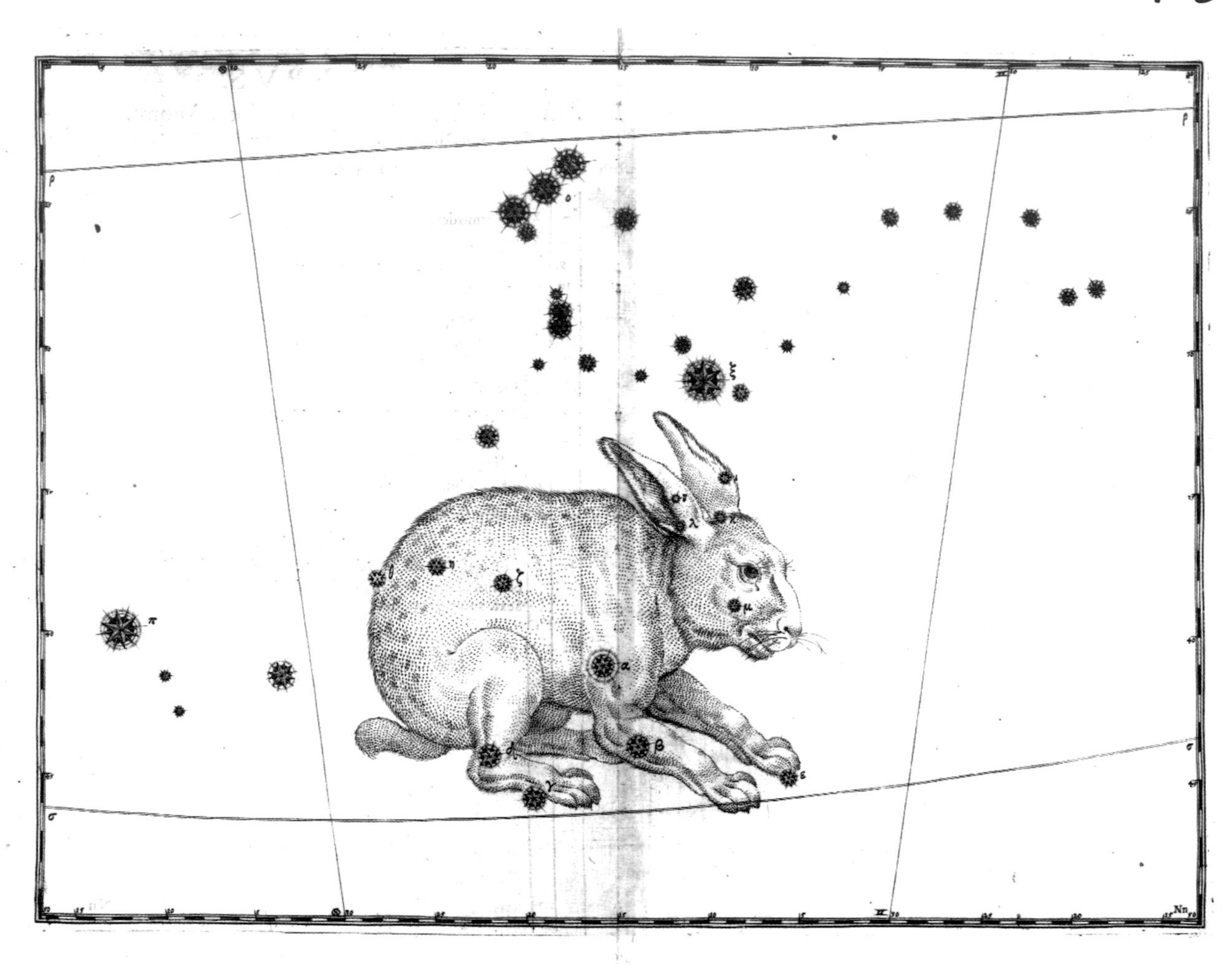

Lepus, the Hare, an ancient but obscure constellation beneath Orion, contains one of the sky's reddest stars, R Leporis. Southern Orion appears at the top. Unknown to Bayer, R Lep is just off the end of the Hare's face, the animal appearing to stare right at it. *(Johannes Bayer. 1603.* Uranometria. *Courtesy of the Rare Book Room and Special Collections Library, University of Illinois.)*

Residence:	LEPUS
Other name:	HIND'S CRIMSON STAR
Class:	N6 CARBON GIANT MIRA
Visual magnitude:	5.5 TO 11.7
Distance:	820 LIGHT YEARS
Absolute visual magnitude:	–1.5 TO 4.7
Significance:	COOL CARBON—AND THE SKY'S REDDEST—NAKED-EYE STAR.

R LEPORIS

Some ancient constellations are less well-known than their "modern" (invented 1600–1800) counterparts. Everyone knows the Southern Cross, but how often do we extol the glories of Lepus, the Hare? But glories there be, including the blazing globular cluster Messier 79. Perhaps best of all is a single, quite spectacular star discovered by J. R. Hind in 1845. Observers still rave over its extraordinary color. Likened to a glowing coal in a black sky, Hind's Crimson Star makes Herschel's Garnet Star (**MU CEPHEI**) seems pale by contrast.

The formal name "R Leporis" implies variability, the Roman letters continued for variables after Bayer's last-used "Q." R Lep, as it is known to its fans, is a long-period variable similar to **MIRA**, changing its brightness by a factor of up to 300 during a long, 430-day period. Like most "Miras," R Lep's variability is itself variable, the star sometimes not making naked-eye visibility at its maximum, while other times reaching an easily accessible fifth magnitude.

To be a "Mira," a star must be a cool giant, and cool stars are reddish, the shade of **BETELGEUSE**, Mu Cephei, or Mira itself. The term giant is certainly apt. From its distance of 820 light years and direct measures of its angular diameter, R Lep would, if put in place of the Sun, stretch 95 percent of the way to Mars. As a Mira-type variable, R Lep is also luminous, its maximum radiation over 300 times the visual solar power (while at minimum plunging only to solar brightness). Such variation is deceptive, though, as cool R Lep, like other "Miras," radiates most of its light in the infrared. As a result, visual variations are caused as much (or more) by changes in temperature as changes in energy output.

Two additional factors contribute to the depth of R Lep's color. First, it has an especially low temperature of only 2050 K, making the star one of the coolest giants known. More important, however, is that R Leporis is an arch "carbon star." When Father Secchi classified the stars 150 years ago, he found two kinds whose absorptions were in the form of broad "bands" rather than narrow "lines." The majority had bands that shaded to the red; the other kind shaded oppositely. The rarer bands were quickly identified with carbon molecules, while the more common bands were later found to be the ubiquitous titanium oxide of the M stars. Harvard's William Pickering, who developed the modern classification system, called the carbon stars class "N" ("C" now used). Carbon molecules, especially cyanogen, absorb a great deal of blue light. As a result, the more carbon, and the cooler the star, the more blue light is absorbed. Carbon stars are therefore quite red. R Lep, partly because of its very low temperature, takes redness to an extreme, making it the sky's riveting "Crimson Star."

The carbon is a by-product of stellar evolution. R Lep, like Mira, has fused its internal core to a mixture of carbon and oxygen and can go no farther. The core is now heating and contracting, while hydrogen fuses into helium, and helium fuses into carbon in shells that embrace the burnt core. Some of the carbon from the helium-burning shell has been lofted to the surface by convection. At one time, surface oxygen dominated carbon. For a brief period when carbon equaled oxygen, R Lep must have appeared as an S star like **CHI CYGNI**. It is now a full-blown carbon star. Injecting the carbon into interstellar space through a stellar wind, R Leporis is preparing to produce a planetary nebula. Most of the carbon in the Galaxy seems to come from stars just like this one. Even your pencil was once part of ancient, long-gone "Crimson Stars."

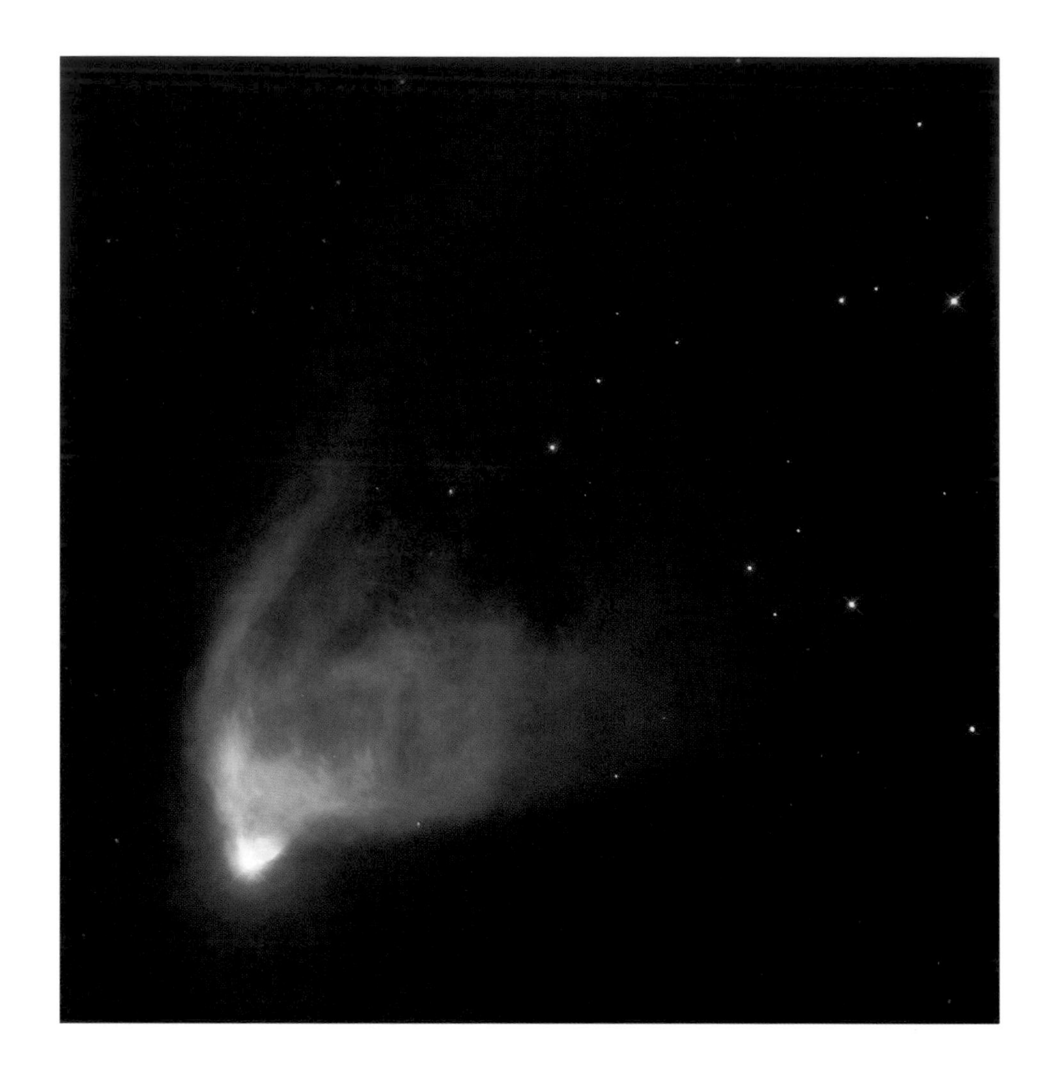

Fan-shaped NGC 2261, Hubble's Variable Nebula, extends northward from the youthful star R Monocerotis. *(NASA and the Hubble Heritage Team [AURA/STScI].)*

Residence:	MONOCEROS
Other name:	HUBBLE'S VARIABLE NEBULA
Class:	AE/BE YOUNG STELLAR OBJECT
Visual magnitude:	10.4 VARIABLE
Distance:	2500 LIGHT YEARS
Absolute visual magnitude:	–4.6?
Significance:	HIGH-MASS PROTOSTAR.

R MONOCEROTIS

When the great 200-inch Hale telescope went into operation at Palomar in 1947, one of the first pictures taken was that of NGC 2261, Hubble's Variable Nebula. In 1916, Edwin Hubble, the famed astronomer after whom the Space Telescope was named, discovered that this fan-shaped reflection nebula was uniquely variable. Large changes—up to four magnitudes—in a star at its apex, R Monocerotis, had been observed since 1861.

The region surrounding R Mon is awesomely complex. Dimmed by 13 magnitudes (a factor of 150,000), the star is so buried in a pocket of dust 200 AU across that it cannot be seen in optical radiation. We know it is there because its light is scattered and reflected from the dust particles embedded in NGC 2261. Shooting out from the deep interior is a bipolar jet moving with a speed over 100 kilometers per second. The main part of Hubble's Variable Nebula extends from the star 1.5 light years to the north, with a faint elongation extending twice as far. Within the nebula is a set of heliacal filaments, suggesting the presence of a twisted magnetic field. A faint counter-fan reaches toward the south, opposite Hubble's Nebula, revealing it to be biconical. Over five light years away to the north along the axis of NGC 2261 lies a cluster of luminous knots, a Herbig-Haro object (HH 39) like those associated with **L 1551 IRS 5**. Small blobs of gas are moving away from R Mon at some 300 kilometers per second. Little imagination is needed to relate them to the inner jets, which speed for light years until they hammer the local interstellar gases, sweeping them up in a shock wave, in a giant sonic boom.

Perpendicular to the biconical nebula is a huge, thick disk of molecular hydrogen, revealed through radio radiation from carbon monoxide, that extends more or less east-west over 2.5 light years from the inner star. The powerful, magnetically focussed flow that originates near R Mon has punched a biconical hole in the extended disk, its sides making the reflection nebula. Some of the variation in Hubble's Nebula reflects the stellar variations, but some is also coming from shadows cast by thick dust moving in the inner structure. Everything points to a star glaringly caught in a state of formation.

R Monocerotis itself is a classic "Herbig Ae/Be star." Made of (obviously) classes A and B, the Ae/Be stars have emission features in their spectra. Associated with dark and bright nebulosity, with thick clouds of interstellar gas and dust, their spectra show that they are simultaneously accreting and losing mass. They are youthful high-mass versions of **T TAURI** stars, and are distinct from aging Be stars like **GAMMA CASSIOPEIAE**, whose mass ejections are tied to their rapid rotations. While T Tauri stars are destined to become mature stars rather like the Sun, the Ae/Be stars, with masses twice solar or more, will become the dwarfs of the upper main sequence.

R Mon's mass is estimated to be about ten times solar. From a 35,000-K surface, it radiates somewhere around 5500 solar luminosities into its dense surroundings. From the inner disk, it accretes about a ten-thousandth of a solar mass per year. That R Mon is truly youthful is confirmed by a 1.5 solar mass classical T Tauri companion (R Mon-B) that orbits at a distance of only 500 AU. R Mon-A, the bright star that causes all the fuss, will someday be a B0 (or so) dwarf, while the companion will be a mid-A star. The pair will eventually become yet another of the Galaxy's myriad of double stars. The inner disk surrounding R Mon-A is crudely the size of our planetary system. Could planets be in the making as well?

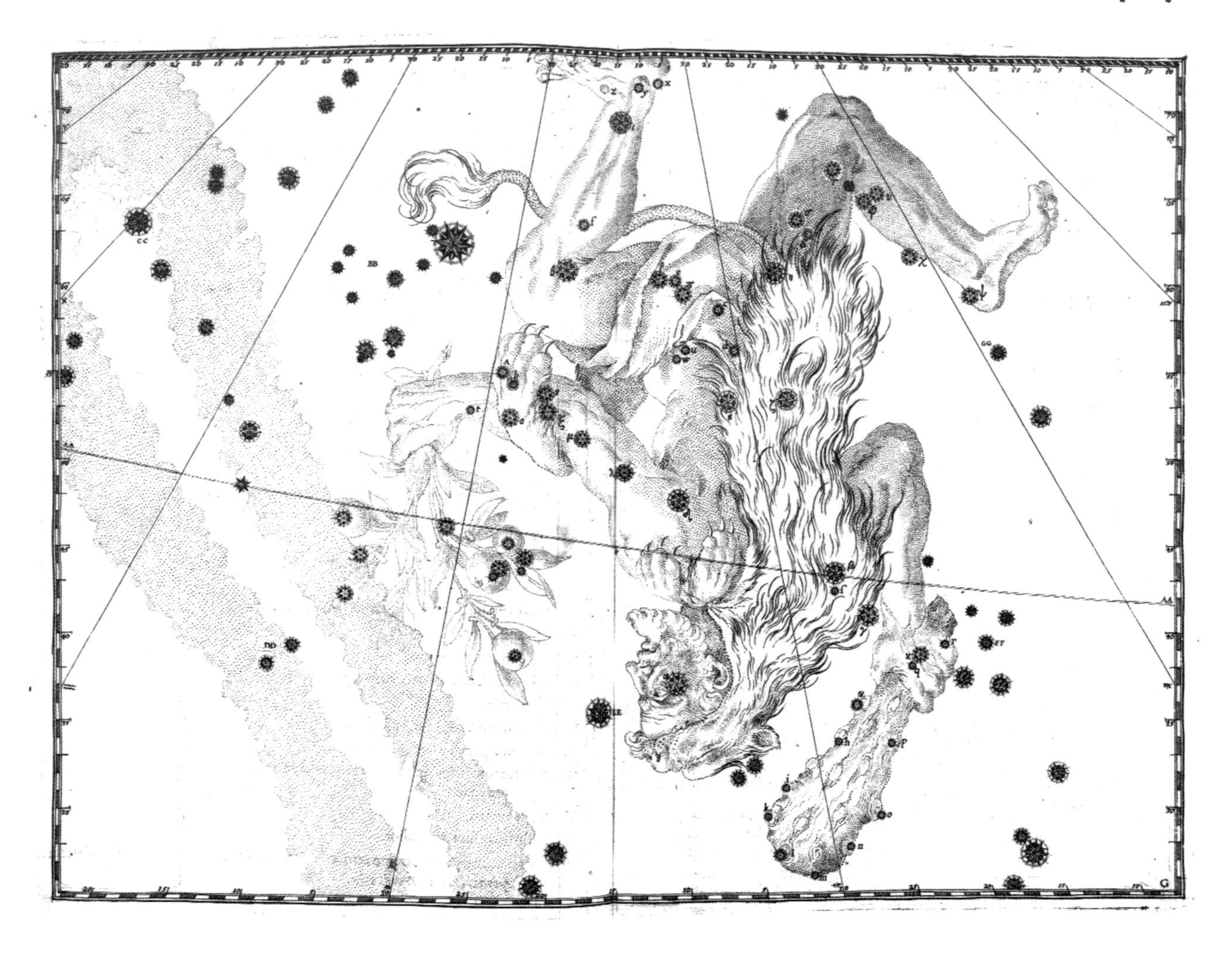

The supergiant star Rasalgethi—Alpha Herculis—appropriately marks giant Hercules's head, the hero traditionally appearing upside down. *(Johannes Bayer 1603.* Uranometria. *Courtesy Rare Book Room and Special Collections Library, University of Illinois.)*

Residence:	HERCULES
Other name:	ALPHA HERCULIS
Class:	M5 SUPERGIANT + (G5 GIANT + F2 DWARF CLOSE BINARY)
Visual magnitude:	3.5 VARIABLE
Distance:	380 LIGHT YEARS
Absolute visual magnitude:	3.48 AND 5.39
Significance:	TRIPLE STAR THAT REVEALS A STELLAR WIND.

RASALGETHI

Among the oldest celestial figures known, Hercules was handed down to the ancient Greeks from even earlier times as the mysterious "Kneeler," shown as the figure of a man seen upside down, his head toward the south. Though only Hercules's fifth-brightest star, Rasalgethi's prominent position as "the Kneeler's Head" (the Arabic meaning of the name) made it Bayer's Alpha star. The accolade is well-deserved.

Third magnitude as seen from Earth, Rasalgethi is a magnificent reddish giant, maybe even a supergiant, with a surface temperature of 3300 K. At a distance of about 400 light years, the star is visually almost 500 times more luminous than the Sun. Factoring in the infrared radiation that pours from the cool surface boosts the total luminosity to 17,000 solar. Rasalgethi is so large and close that its angular diameter can be rather easily measured at 0.032 seconds of arc, which gives it an impressive physical diameter 400 times that of the Sun (3.8 AU), or 25 percent larger than Mars's orbit. Well into advanced evolution, this star of around 7 or 8 solar masses is probably fusing helium into carbon in its deep core. As are so many huge stars, Rasalgethi is a relatively erratic "semiregular variable." It wanders between magnitudes 3 and 4 with ill-determined periods of 50 and 130 days superimposed on one of six years.

However, the star's proportions are not what make it special, as the sky contains many stars that are equally grand. Zero in on it with a telescope, and you see it as a double, a white fifth magnitude (5.39) companion glowing only 4.6 seconds of arc away. Contrast effects make the pair remarkably lovely, observers reporting "orange and emerald green." Separated by a distance of at least 550 AU, the two take more than 3500 years to make a full orbit. Again, however, many stars are double, and many are at least as pretty. So there must still be something else.

The companion is also double, making Rasalgethi a triple star. The components of Rasalgethi B, however, are not separable by eye. Instead, we see that the spectrum is a composite of a 4-solar-mass G5 giant coupled with a 2.5-solar-mass F2 dwarf. Responding to the Doppler effect, the two sets of absorptions in the spectrum shift back and forth over an orbital period of 51.578 days. The pair is separated by only 0.4 AU, a gap about the size of Mercury's orbit. From the red M5 supergiant, Rasalgethi A, the two would appear as two bright points up to 3 minutes of arc apart, just resolvable with the human eye. From the whirling pair, the supergiant would appear nearly ten times bigger!

That still is not the special part. The tightly orbiting pair provides a background that told us a great deal about how stars work. In the spectrum of Rasalgethi B lies a set of absorptions that do not shift back and forth and, therefore, cannot belong to either of the stars themselves. The same absorptions show up in the spectrum of the supergiant. There, however, they are Doppler-shifted to shorter wavelengths, showing gas coming at us at a speed of about 10 kilometers per second. Together, the observations reveal not only that the supergiant has an expanding atmosphere or wind, but that this wind entirely encompasses the binary companion over 500 AU away. Hercules' great supergiant and its lowly companions thus provided some of the first evidence that stars lose great quantities of mass as they die. Even now, Rasalgethi A is whittling itself down to become a massive white dwarf. The G giant of the close pair will shortly follow, making the F dwarf as king of the system until it too dies away in similar fashion, leaving a triple white dwarf behind.

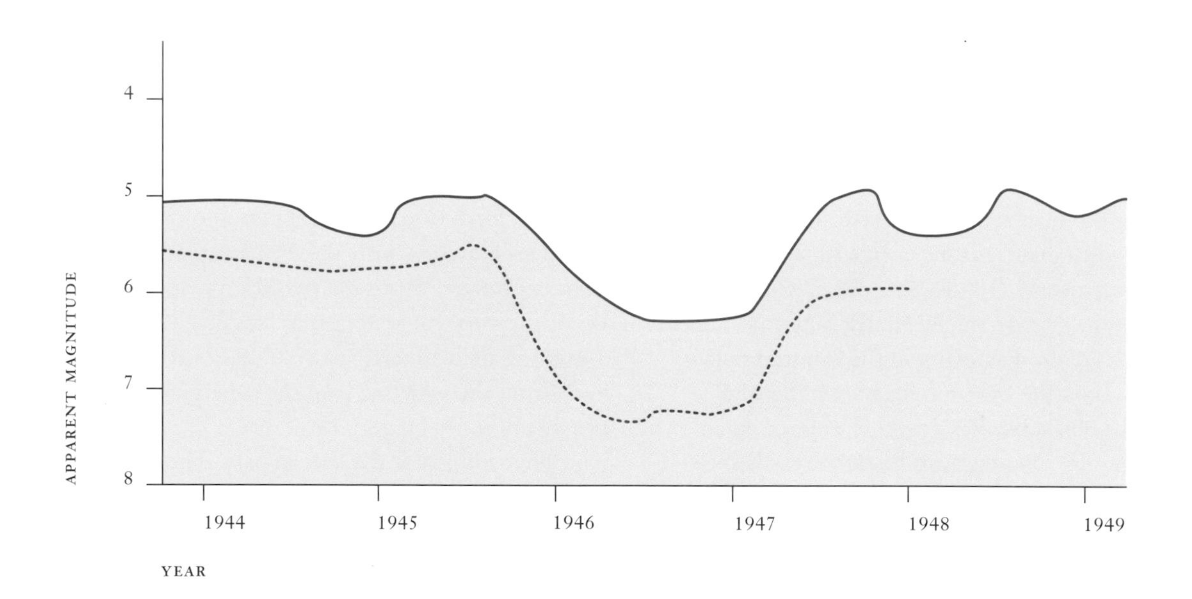

Shortly after GAMMA CASSIOPEIAE faded back from its 1937 outburst, hypergiant Rho Cassiopeiae (see #95) faded, dimming by about one visual magnitude in 1946 (top), and even more when seen in blue light (bottom). *(Adapted from W. F. Beardsly, from a paper in the* Astrophysical Journal Supplements 5 [1961]: 381. *Visual observations courtesy of the AAVSO.)*

Residence:	CASSIOPEIA
Other name:	7 CASSIOPEIAE
Class:	F8 HYPERGIANT
Visual magnitude:	4.5 VARIABLE
Distance:	8000 LIGHT YEARS
Absolute visual magnitude:	–9.6
Significance:	ERRATICALLY DIMMING HYPERGIANT.

RHO CASSIOPEIAE

Summer 1946. Less than a year after the conclusion of World War II, few had reason to notice a fourth magnitude star in Cassiopeia. Within that bright figure set into a rich part of the Milky Way, the star is nearly lost anyway, going only by the lowly Greek letter name "Rho." Most stars that "do" something brighten, but Rho Cas did just the opposite.

But we are ahead of our story. Years ago, stars were first divided into ordinary dwarfs like the Sun and into bigger ones, the giants. A set of still larger stars could only be called supergiants. The differences involve more than physical size: for a given temperature, giants are far more luminous, and the supergiants even more so. What do you call stars when they are larger and brighter than supergiants? Hypergiants, of course. The class was first needed for ultraluminous stars in the Magellanic Clouds, our small satellite galaxies. We then found them in the Milky Way as well. They are a rare breed (perhaps only a dozen are known), and include **ETA CARINAE**, **P CYGNI**, and Rho Cassiopeiae. (Nothing, however, is without argument. Rho Cas is also referred to as a bright supergiant, but a rose by any other name. . . .) Estimated to be 8000 light years away and dimmed by two magnitudes by interstellar dust, it still shines at fourth magnitude, radiating 550,000 times more light than the Sun.

Even among the hypergiants, stars like Rho Cas are rare. Neither red (cool) nor blue (hot) like most, Rho Cas lies in between, a yellow hypergiant. With a temperature of 7300 K, its energy pours out in the visible part of the spectrum. The temperature and luminosity tell of a distended surface 450 times larger than the Sun, or 4.3 AU, which is 40 percent larger than the Martian orbit. Rotating at least at 29 kilometers per second, the star could take up to two years to make a full spin. Though it has no companion from which to gauge its mass, the immense luminosity suggests roughly 40 times solar.

Theory shows that hydrogen-fusing dwarf stars from 10 to about 60 solar masses evolve from class O first to become blue supergiants and then into red class M supergiants. However, from around 40 to 60, they loop back, turning from red supergiants back into much hotter and smaller blue supergiants. Higher than 60, they stay as blue supergiants. Rho Cas now seems to be on its way back from being a red supergiant, when it must have been five times larger. If so, it is bouncing against the "yellow evolutionary void" in which stars become unstable and do not like to linger. And Rho Cas certainly is unstable. It is an irregular variable, or at best as a semi-regular, and seems to have multiple periods of 820, 350, 510, and 645 days. But these change, so the star may really be quite unpredictable.

But back to the summer of '46. Rho Cas took a dive from fourth to sixth magnitude and, more remarkably, altered its spectral class. Pumping a huge amount of gas into an expanding thick atmosphere, it chilled from class G or F to cool class M. Less than a year later, it was returning to normal, though still only back at class G8. The star did not so much dim as cool, the lower temperature causing it to place much of its radiation in the invisible infrared. In its more stable state, Rho Cas still blows a 10-kilometer-per-second wind at a rate of a hundred thousandth of a solar mass per year—100 million times the flow rate of the solar wind. It does not have much time left before it grows its iron core and blossoms into the sky as a stunning supernova.

Long lanes of dark star-forming dust invade the stars of the Milky Way, the darkest pointing to the right directly at Rho Ophiuchi. Below Rho Ophiuchi lies ANTARES (also see #8), the bright star of Scorpius. *(© Anglo-Australian Observatory, photograph from UK Schmidt plates by David Malin.)*

Residence:	OPHIUCHUS
Other name:	5 OPHIUCHI
Class:	B2 SUBGIANT + B2 DWARF
Visual magnitude:	5.02 AND 5.92 (4.63 COMBINED)
Distance:	395 LIGHT YEARS
Absolute visual magnitude:	–2.54 AND –1.64
Significance:	A MARKER FOR DARK CLOUDS AND INTENSE STAR FORMATION.

RHO OPHIUCHI

Can't get into the Southern Hemisphere to see **ACRUX**? Try its more northerly clone, Rho Ophiuchi. Acrux consists of a B0.5 subgiant set 4.4 seconds of arc from a B1 dwarf, while Rho Oph is made of a pair of B2 stars 3.2 seconds of an arc apart. The biggest difference is brightness. Rho Oph's combined pair shines at fourth magnitude instead of first because of its somewhat greater distance, lower stellar temperatures, and the absorption of starlight by interstellar dust. Separated by a minimum distance of 400 AU, the Rho Oph pair takes at least 2000 years to orbit. With temperatures of 22,400 K, they emit most of their light in the ultraviolet, the brighter radiating 4900 solar luminosities, the fainter 2100, implying respective masses of 9 and 8 solar. Though the brighter is classed a subgiant, both are probably still hydrogen-fusers. Rotation velocities are very high, around 300 kilometers per second, and the stars may be on their way to becoming mass-losing emission line stars like **GAMMA CASSIOPEIAE**.

None of these characteristics makes the stars particularly special. Rho Oph, however, is set not only in the Milky Way (like Acrux), but within an extraordinary complex of bright and dark nebulae. Say "Rho Ophiuchi," and what comes to mind is not the star but the "Rho Ophiuchi dark cloud." Located just north of **ANTARES** in Scorpius, Rho is surrounded by a blue reflection nebula (IC 4604) that, like the **MEROPE** nebula, glows from starlight scattered by dust grains.

IC 4604, ten light years across, is but the illuminated portion of a nest of thick, dark, dusty clouds. So much dust is in the line of sight to Rho Oph that its light is dimmed by a factor of six, remarkable for a star so close to us. Were the area free of dust, Rho Oph would appear nearly second magnitude. Moreover, the dust is weird. Interstellar grains, usually no more than one thousandth of a millimeter across, absorb and scatter blue light far better than they do red. As a result, bright reflection nebulae are commonly blue, and the stars behind dust clouds are redder than they should on the basis of their temperatures. The amount of dust absorption is estimated by a standard formula from a star's apparent reddening. The dust toward Rho Oph absorbs more efficiently than expected, the result of dust grains that are larger than normal. No one knows why.

Around Rho's reflection nebula, the sky turns dusty black. Extending to the east is a remarkable set of streamers up to 13 degrees (over 100 light years) long. The contrast of blue, red, and dark nebulae, all backed by the Milky Way's millions of stars, is stunning. The dark clouds are chilled by the dust, which blocks starlight that would otherwise heat them. The cooling results in the formation of molecules that include carbon dioxide, water, and many that are much more complex. Within these molecular clouds, there are denser regions where stars are forming. Radio and infrared observations that penetrate the dust reveal everything from dense knots that are just beginning to contract into stars under the force of gravity, through "protostars" where the formation process is well underway, to active young **T TAURI** stars and all their associated phenomena. Particularly busy is a dark region on the eastern edge and perhaps 50 light years in the back of the Rho Oph reflection nebula. Within a volume only a few light years across, there are dozens of stars that are "works in progress." Though your eye cannot see into the clouds, you can at least admire visible Rho Oph and know what is going on deep within its black heart.

The globular cluster Messier 3 in Canes Venatici (see also #23) is loaded with RR Lyrae stars. *(NOAO/AURA/NSF.)*

Residence:	LYRA
Other name:	HD 182989
Class:	F5 HORIZONTAL BRANCH
Visual magnitude:	7.8 VARIABLE
Distance:	750 LIGHT YEARS
Absolute visual magnitude:	1.0
Significance:	THE PROTOTYPE RR LYRAE PULSATING VARIABLE.

RR LYRAE

Pulsating Cepheid variables, epitomized by naked-eye **DELTA CEPHEI**, are among the most important of the different kinds of stars. Because they are easily-visible supergiants whose absolute magnitudes tightly correlate with their periods (which run from a few days up to 50), they collectively provide a "standard candle" with which to find distances to other galaxies. At the short-period end, they seem to merge with under-a-day pulsators epitomized by eighth magnitude RR Lyrae. Though there are physical similarities, the two kinds of stars are remarkably different.

RR Lyrae is a helium-burning giant that varies as precisely as a well-oiled clock between magnitudes 7.2 and 8.0 with a period of 13 hours, 36 minutes, and 14.9 seconds. From its distance of 750 light years, it shines with an average luminosity 40 times that of the Sun and a surface temperature of 6700 K, roughly appropriate for a mid-F star. Traditional spectral classification, which is based on stars of normal solar chemical composition, is a bit problematic, however. RR Lyrae is highly deficient in metals, having only 4 percent of the Sun's iron content, which changes the appearance of the spectrum and makes comparison with established standards rather difficult. The low heavy-element content provides a clue to the star's nature. So does the velocity (relative to the Sun) of 240 kilometers per second. Together, they reveal that RR Lyrae is not one of the stars of the Galactic disk, but is an interloper from the low-metal Galactic halo like **KAPTEYN'S STAR**. Though RR Lyrae stars are scattered here and there throughout the halo, their real home lies within the massive globular clusters, huge ancient assemblies that can contain millions of stars. Some globulars have RR Lyrae stars, others do not, but so many do that "RR Lyraes" are also called "cluster variables."

Globular clusters are so old (12 to 13 billion years) that they contain no hydrogen-fusing dwarfs above 0.8 solar masses. These, just below one solar mass, are the RR Lyrae progenitors. If they had the Sun's metal content, the evolving stars would develop into cool class K giants with luminosities about 50 times solar. Drop the metal content, however, and they become more hot, blue, and transparent. Add small mass variations, and the set spreads out in temperature from class K to B on the Hertzsprung-Russell diagram into a "horizontal branch." In the middle of this range, in classes F and A, the stars become unstable, pulsating RR Lyrae stars.

Since RR Lyrae stars all have about the same absolute visual magnitudes, they provide excellent standard candles for distance measurement. Yet for all their significance, these variables are still not fully understood. They come in two flavors, or "frequency modes." An organ pipe or guitar string will vibrate with a fundamental frequency, but also with a variety of higher -frequency overtones. RR Lyrae stars with longer periods of over half a day vary with their natural fundamental frequencies, while those with shorter periods are in their "first overtones" (rather like **POLARIS**). While the overtone pulsators are quite stable, the magnitude ranges of the "fundamentals" slowly change over periods of a few weeks, RR Lyrae—a fundamental pulsator—taking 41 days. We have no idea why.

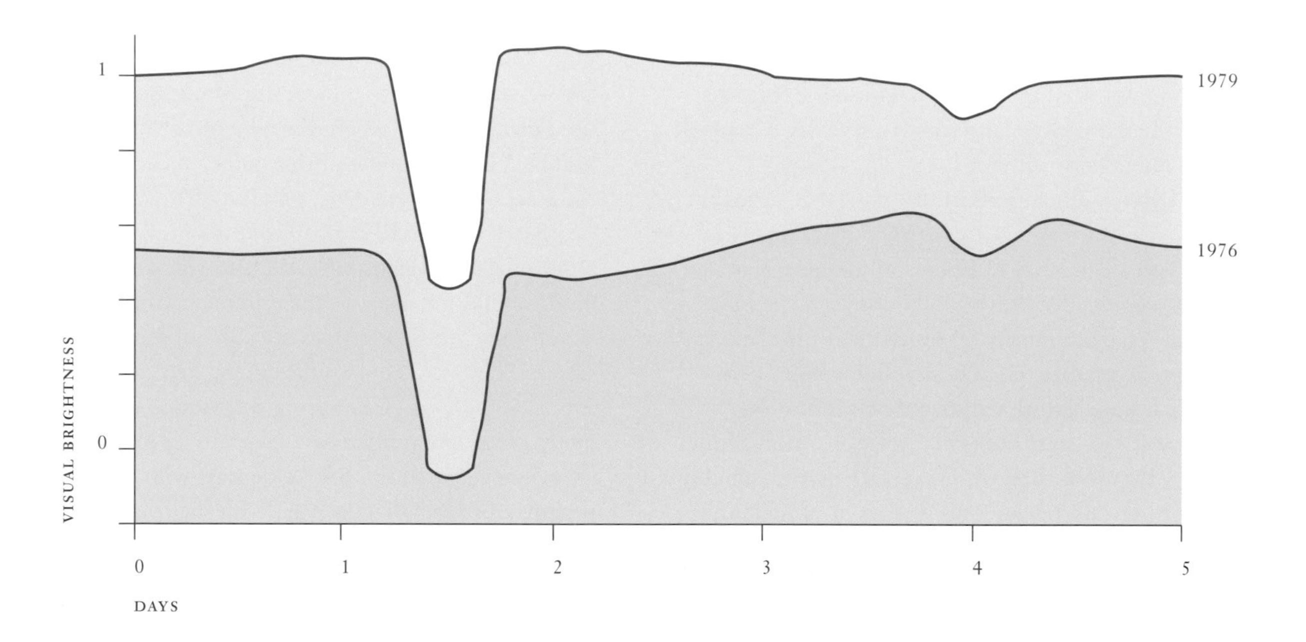

The mutual eclipses of RS Canum Venaticorum are evident as deep drops in light. The variations while the two are not eclipsing are caused by starspots moving in and out of the field of view. The scale is set for the top curve. The lower curve is dropped for comparison. *(Adapted from Y. W. Kang and R. E. Wilson, from the* Astronomical Journal 97 [1989]: 848.*)*

Residence:	CANES VENATICI
Other name:	HD 114519
Class:	F5 DWARF + K0 SUBGIANT
Visual magnitude:	8.2 TO 9.3
Distance:	350 LIGHT YEARS
Absolute visual magnitude:	3.06 COMBINED
Significance:	HEAVILY SPOTTED SYNCHRONOUSLY ROTATING DOUBLE STAR.

RS CANUM VENATICORUM

If the orbital plane of a double star lies along the line of sight, the stars will eclipse, the most famed eclipser being **ALGOL**. However, a good many eclipsing doubles also vary outside of eclipse: matter can flow from one star to another, forming a gaseous stream whose brightness depends on how we view it. The stars may be tidally distorted like eggs, and as they orbit we see them alternately side-on and brighter, then end-on and fainter. One star's light may also be reflected from the other, the brightness of the pair then depending on how the stars align relative to Earth.

RS Canum Venaticorum gave a whole new look to such variations. It is made of a mid-temperature F4 (6800 K) hydrogen-fusing dwarf and an evolving class K0 (4580 K) subgiant with a dead helium core, which together are five times more luminous than the Sun. While the subgiant is twice the size of the dwarf, its coolness causes it to radiate less energy, so the two stars have about equal luminosities. Since the K star radiates more of its light in the infrared, the F star is visually the brighter. The spectrum shows strong emissions from calcium and hydrogen. Every 4.797851 days, the subgiant orbits in front of the dwarf and blocks it completely, dimming the double by a full magnitude. In between, the dwarf cuts out some of the light of the subgiant. Brightness changes outside of eclipse are not explainable by ordinary means, however. At least not until you think about the Sun.

Looking at a solar photograph, the eye is immediately drawn to the dark sunspots. The outer third of the Sun is in a state of roiling convection which, with solar rotation, generates a magnetic field that gets powerfully amplified at the base of the convection zone. Ropes of intense magnetism float to the stellar surface, where they inhibit the convection and create the spots that chaotically come and go. Spots are related to bright flares and to spectral emissions from the chromosphere (the thin layer just above the stellar surface) that show up in most solar-type stars. Since rotation and convection produce the magnetism that creates the activity and spots, rapidly rotating stars have greater activity.

The components of RS Canum Venaticorum are so close that they raise significant tides in each other. The tides raised by the Moon on Earth dissipate rotational energy and slow the our planet; tides raised by the Earth on the Moon have stopped its rotation altogether, locking it on to us. The same thing has happened to RS Canum. Its stars' rotational periods match the orbital period, causing them to rotate five times more quickly than the Sun. The cooler star, which has begun to evolve to be a red giant, has developed a deep convection zone. This zone and the quick spin raise huge starspots that can cover much of the stellar surface. As the spot or spots rotate in and out of view, the star dims or brightens by a few tenths of a magnitude.

RS Canum is the prototype for a great many similar stars. They start as class F or so dwarfs of almost equal mass. The one with the greater mass evolves first and expands while its internal hydrogen fusion shuts down. The stellar rotations synchronize with orbital revolution, and the new subgiant develops the spots and chromospheric emissions. Many of these stars are highly active, kicking off huge flares that are visible from X-ray to radio wavelengths and that dwarf those of the Sun. As the subgiant continues to expand, it can absorb its mate. The orbital energy goes into rotational energy, and a single spotted **FK COMAE BERENICES** star may be born from a marriage literally made in heaven.

RS Ophiuchi (seen "at rest" at the top) had one of its great recurrent outbursts in 1967 (bottom). The out-of-roundness of the image is an artifact of the telescope. *(Lowell Observatory photographs.)*

Residence:	OPHIUCHUS
Other name:	NOVA OPHIUCHI
Class:	M2 GIANT + WHITE DWARF
Visual magnitude:	11.5 MINIMUM
Distance:	2000 TO 5000 LIGHT YEARS
Absolute visual magnitude:	2.6 TO –1.7
Significance:	THE BEST-KNOWN RECURRENT NOVA.

RS OPHIUCHI

Most novae have a year attached to them, for example Nova Herculis 1934 (**DQ HERCULIS**) or **NOVA CYGNI 1975**. Applying this rule, RS Ophiuchi becomes Nova Ophiuchi 1898-1933-1958-1967-1985, a bit long to use. Unlike most novae, RS Oph is a recurrent nova that unpredictably pops off every 20 years or so. In 1985, it brightened in just a few days from nearly magnitude 12 to 5th, with the explosion radiating from X-ray to radio wavelengths. Within two days, the star disap peared from naked-eye visibility and within 200 days was back to normal. Previous outbursts were similar (as have been those of several others of the breed). The distance is very uncertain, estimates ranging from 2000 to 5000 light years. At peak, however, RS Oph could have been as luminous (counting radiation all along the spectrum) as 37,000 Suns.

Ordinary novae like DQ Herculis and Nova Cygni 1975 are produced by short-period double star systems, wherein a low-mass dwarf is so tidally distorted that it dumps fresh hydrogen onto the surface of a dense white dwarf. When the pressure and temperature of the compressed hydrogen is great enough, it explodes in a thermonuclear runaway. After decades, the white dwarf returns to normal.

Like the typical nova candidate, RS Oph is a double star that includes a hot white dwarf. However, the primary star—the one visible through the telescope—is an evolved, helium-fusing red giant. While the DQ Herculis pair takes only four and a half hours to complete an orbit, the RS Oph team takes 460 days. The red giant shines with perhaps 1000 solar luminosities, the white dwarf (and its accreting matter) with well over 100. The red giant is tidally distorted and passes mass onto the white dwarf through its wind.

Since high-mass stars are rare, so are high-mass white dwarfs. The great majority of novae are produced by low mass white dwarfs. The time for hydrogen build-up to produce an ordinary nova may be hundreds of thousands of years, so we get to see but one explosion. However, if the white dwarf is massive, it will have a much higher gravitational field. The conditions for sudden fusion will be reached in a much shorter time, resulting in detonations every few decades instead of every few hundred millennia. RS Oph's white dwarf is estimated to have a mass just shy of the allowed limit of 1.4 solar. It takes only a couple millionths of a solar mass of hydrogen to produce the outburst. As energetic as the explosion is, however, not all of the infalling matter gets blown off: about 10 percent remains behind. As a result, in spite of the repeated detonations, the RS Oph white dwarf continues to grow. Eventually it will hit the white-dwarf limit and explode.

If the theory is correct, the two stars are conspiring to produce a Type Ia supernova like **TYCHO'S STAR**. The white dwarf's mass is uncertain. Orbital solutions actually show the white dwarf curiously beyond the allowed limit, so it could go anytime: maybe even tonight.

83

Gamma Rays (Ulysses Spacecraft)

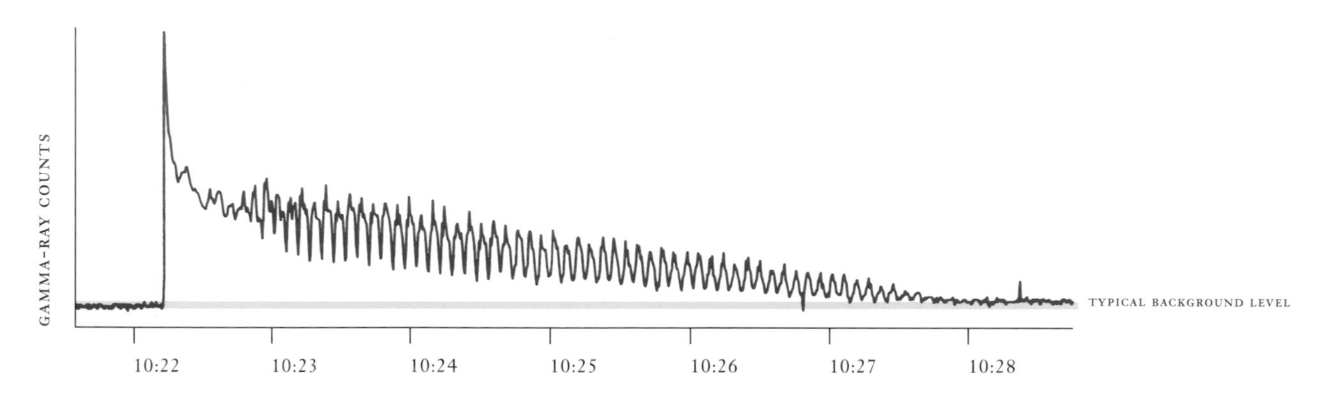

Reflected Radio Waves (Colorado, USA)

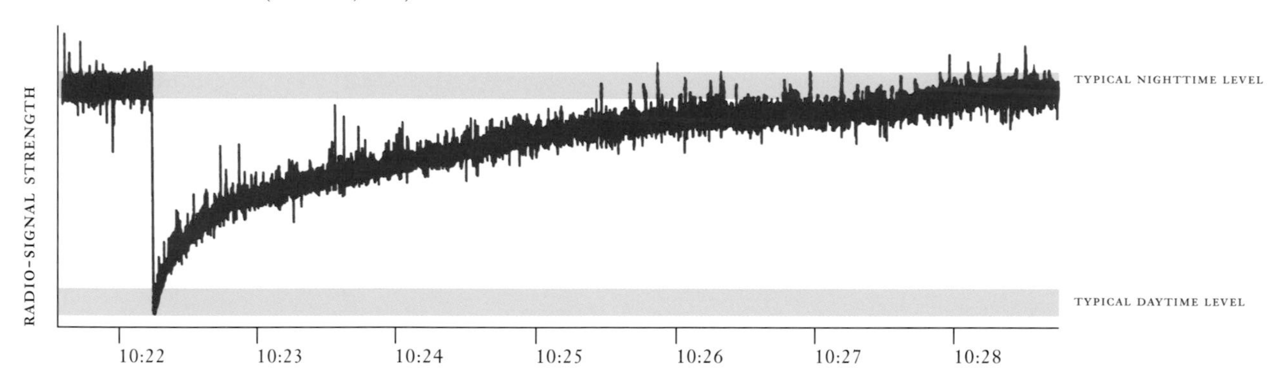

UNIVERSAL GREENWICH TIME (AUGUST 27 1998)

In its great outburst of 1998, SGR 1900+14 rained a spectacular storm of gamma rays upon the Earth (top). The resulting thickening of the Earth's ionosphere (the high-altitude layer where the air is ionized), is seen in the dramatic weakening of reflected radio waves. *(Courtesy of Michael Johnson, adapted with permission from Umran Inan, both of Stanford University, and* Sky and Telescope *[January 1999]: 22.)*

Residence: AQUILA
Other name: GB 790324
Class: SOFT GAMMA-RAY REPEATER
Visual magnitude: NONE
Distance: 20,000 LIGHT YEARS
Absolute visual magnitude: NONE
Significance: ULTRA-MAGNETIC NEUTRON STAR THAT "ADJUSTED" ITSELF AND IONIZED THE EARTH'S ATMOSPHERE.

SGR 1900+14

Astronomical superlatives abound: stars the size of the Solar System, small as a village, **ETA CARINAE** pouring 5 million solar luminosities, brown dwarfs so faint and cool the Sun could have one as a companion and we would not know it, X-ray-generating black hole binaries. . . . What next?

"Soft gamma-ray repeaters"—SGRs—emit machine-gun-like bursts of relatively low-energy gamma rays that individually last about a hundredth of a second. Since gamma rays are not readily analyzable from the ground, only spacecraft can easily detect and study them. SGRs are extremely rare, and only three are known in our Galaxy. Two, SGR 1806–20 in Sagittarius and SGR 1900+14 (the numbers are coordinate positions), are particularly well studied. Both were found to be aligned with distant supernova remnants, and it seemed a good bet that the SGRs were some kind of neutron star. The wager was won through X-ray observations, those from SGR 1806–20 and 1900+14, respectively, carrying pulses with rather long periods of 7.47 and 5.16 seconds. The SGRs are not only neutron stars, they are radio-quiet pulsars like **GEMINGA**. But the comparison stops there.

In 1992, theoreticians suggested that, if conditions are right, an iron core collapse could make a supernova, and this ultrapowerful, convective dynamo could create a neutron star with a magnetic field far beyond anything yet observed. Such stars, with magnetic fields 100 or more times that found in an ordinary pulsar, are now called "magnetars." The intensely beamed radiation from pulsars and the winds from their surfaces, the latter moving at close to the speed of light, remove energy from the spinning stars and slow them. The more powerful the magnetic fields, the greater the rates of slowdown. As a result, a pulsar's spindown rate, combined with the pulse (rotation) period, gives both the magnetic field strength and the pulsar's age. The two SGRs are found to have magnetic fields over 500 trillion times that of Earth! As far as we know, they carry the strongest magnetic fields in the Universe. Their ages are estimated at about 10,000 years, which accords well with the ages of the associated supernova remnants. These soft gamma-ray repeaters must be the predicted magnetars. Given the average spin-down rate and the number we see, about one should be created in a supernova blast every 1000 years, a tenth or twentieth the creation rate of normal pulsars. Most likely, magnetars are the progeny of the rarest and highest-mass stars.

A newly formed pulsar, bulging at the equator because of its rotation, is so dense that it develops a solid crust that re-adjusts as the spin slows. In the resulting starquake, the magnetic fields also re-adjust like a gigantic solar flare, releasing immense energy. The bursts of power can be awesome. On August 27, 1998, at 10:22 Greenwich time, SGR 1900+14 blasted 5 billion times as much energy as the Sun radiates each second. The giant outburst was so powerful that it saturated the electronics aboard orbiting spacecraft. The effect it had on our atmosphere was even more impressive. About 60 kilometers above the Earth, the air is ionized by sunlight, producing an ionosphere responsible for short-wave radio communication. At night, the ionosphere rises to 85 kilometers. The outburst from SGR 1900+14 drove the ionosphere from its standard nighttime level to its standard daytime level. SGR 1900+14 is 20,000 light years away! Imagine what might have happened if the magnetar had been 100 light years distant—we were hit with 40,000 times as much energy. Stars are not just celestial points of light; they can reach out to touch us.

84

Sigma Octantis, the star nearest the center of rotation, rounds the south celestial pole. *(© Greg Dimijian.)*

Residence:	OCTANS
Other name:	POLARIS AUSTRALIS
Class:	F0 GIANT
Visual magnitude:	5.47
Distance:	270 LIGHT YEARS
Absolute visual magnitude:	0.88
Significance:	SOUTH POLE STAR.

SIGMA OCTANTIS

Northerners seem almost vainly proud of their Pole Star, **POLARIS**, as if Nature put it there, at (rather, near) the north celestial pole just for them. Southerners, they note, have no Pole Star. They may mumble grudgingly that, yes, there is one, though it is no more than dim Sigma Octantis, way down the Greek letter list in a constellation that is faint to begin with. Even the brightest star of Octans (the Octant), of all things Nu Octantis (not much attention paid to relative brightness here), is but fourth magnitude (3.76), Alpha Octantis a measly fifth (5.15). Skating the border of sixth, at magnitude 5.47, is Sigma, so faint that moonlight or any lights from town will wash it out. At least it is within the confines of a navigational instrument rather than within a Bear (Polaris also Alpha Ursae Minoris). Other than its dim status, however, Sigma Octantis is really quite a good Pole Star. Now only 1.05 degrees off the south celestial pole (just a bit more than Polaris's polar distance of 0.74 degrees), it rightly earns the name "Polaris Australis," first applied in the 1700s.

Physically, the south Pole Star is a mid-temperature (about 7000 K) F0 giant star. From a distance of 270 light years, it shines with a luminosity 34 times that of the Sun, its mass and radius respectively two and four times solar. As a giant that is not very far off the main sequence, it has probably just given up hydrogen fusion or is about to do so. The star is now on its way to becoming first a much brighter red giant, then a helium-fusing K giant, and finally a white dwarf. If it were a giant star now, you would have no trouble at all locating the Pole! As a recently evolved star, it is also among the company of slightly variable "Delta Scuti" stars, a group that is led in brightness by **BETA CASSIOPEIAE**.

Sigma Octantis is also above, both brighter and hotter than, the dividing line that marks vastly increased stellar rotation. Lower-mass stars like the Sun have their outer layers in a state of convection that, along with rotation, creates a magnetic field. The solar wind drags the magnetic field outward with it. The end of the field, however, remains clamped to the Sun. As a result, the Sun and other solar-type stars are magnetically braked so that the older ones naturally rotate slowly (the Sun only once a month). But at higher masses and temperatures (Sigma Octantis beginning life as a hot-end A0 star), the convection layer disappears, and stars rotate more quickly. Polaris Australis is spinning with an equatorial speed of at least 108 kilometers per second (we do not know the orientation of the spin axis), giving it a rotation period of about one day.

But take heart, southerners. Because of the precessional wobble of the Earth's axis, no Pole Star is permanent. A century ago, Sigma Octantis was a better pole star than Polaris (beating it by the same degree Polaris bests Sigma today). Now moving away from Polaris Australis, the south celestial pole will someday enter the rich confines of Carina and Vela, the hull and sail of Jason's ship Argo. For the next 8000 years, the south pole star will keep getting better and better. In AD 5700, the pole will pass close to third magnitude Omega Carinae, then, in AD 7900, to second magnitude Aspidiske (Iota Carinae). Still improving, around AD 9100, the pole will nail Delta Carinae, which, at magnitude 1.96, shines brighter than Polaris. Even better, the South Pole will pass only about 2 degrees from near-first-magnitude (1.78) **GAMMA-2 VELORUM** (with the only Wolf-Rayet star ever to hit either pole) in AD10,600. It is clearly worth the wait.

At left, the white dwarf Sirius B is tucked in next to brilliant Sirius A (see also #2). In a picture taken in X-rays by the orbiting Chandra X-Ray Observatory (right), however, the view is reversed: Sirius B is now dominant because of its much higher temperature. The spikes on the images are artifacts of the telescopes. *(Left: McDonald Observatory; right: NASA/SAO/CXC.)*

Residence:	CANIS MAJOR
Other name:	ALPHA CANIS MAJORIS; THE DOG STAR
Class:	A1 DWARF + DA WHITE DWARF
Visual magnitude:	–1.46
Distance:	8.6 LIGHT YEARS
Absolute visual magnitude:	1.43
Significance:	THE APPARENTLY BRIGHTEST STAR AND COMPANION TO THE BEST-KNOWN WHITE DWARF.

SIRIUS

As light from Sirius enters Earth's atmosphere, it begins to bend, then plunges into a roller coaster of a refractive ride through blowing cells of warm and cold air that break it into diamonds' colors just before it enters the wondering eye of a girl who at that moment knows she will be an astronomer. All stars twinkle, but none like Sirius, the sky's brightest star, one whose white light plays a rainbow. The Greek name, appropriate for a brilliant star of nearly –2 magnitude, means "searing," "scorching," "sparkling."

As the Alpha star of Canis Major, Orion's bigger Hunting Dog, Sirius upstages the constellation as the "Dog Star." Steeped in lore, the first glimpse of the star in morning twilight foretold the rising of the Nile in Egypt's Old Kingdom, whose symbol was a dog. The star now appears later in the year because of precession, the 26,000-year wobble in the Earth's axis that both moves the equinoxes through the zodiac and changes the direction of the celestial pole. On a wider scale, Sirius is the southern apex of the Winter Triangle, which includes Procyon of Canis Minor (the Smaller Dog) and **BETELGEUSE**.

Like **VEGA**, Sirius is an archetypal white (9400 K) class A star, a main sequence, hydrogen-fusing dwarf about twice as massive and 23 times more luminous than the Sun. It appears bright to us mostly because it is only 8.6 light years away. Like many of its class, Sirius displays chemical peculiarities. It is a mild "metallic-line" star, with some metals enhanced at the expense of others as a result of chemical-element separation in a quiet atmosphere. In such a setting, some elements fall under the action of gravity, while others are lofted via the pressure of absorbed radiation.

Sirius's greatest claim to fame may be its dim companion. All stars move relative to others as they orbit the Galaxy, the constellations dissolving over time. Most stars move in straight lines against the distant background. But Sirius and Procyon drunkenly wobble, the result of orbiting companions respectively seen in 1854 and 1895. Sirius's companion, Sirius B, is 10,000 times fainter than Sirius proper. It is also blue-white, with a temperature of 27,000 K. To be that hot but so dim requires a tiny radiating area, a diameter only two-thirds that of Earth. Though the companion to **40 (OMICRON-2) ERIDANI** was the first to claim the name, Sirius B is certainly the most famed of all white dwarfs.

The most amazing aspect of Sirius B and other white dwarfs is not size but mass. Analysis of the double-star orbit shows it to be as heavy as the Sun. All that mass packed into a sphere the size of Earth requires an average density of 1 metric ton per cubic centimeter. A billiard ball made of the stuff would weigh 70 tons, about the same as an army tank. The density is so great that heavier helium atoms sink out of sight, giving Sirius B a pure hydrogen surface layer. It is the archetypical "DA" white dwarf, as opposed to helium-rich "DB" white dwarfs like **HZ 21**.

Such high density is actually no problem, as the atom is almost all empty space, only 1 part in 1000 trillion taken up with "matter." (What you feel as "solidity" involves electrical forces.) Squeezed down by only 1 million in volume, even white dwarfs seem vacuous. The squeezer is gravity. As the end products of solar-type stars, all thermonuclear fusion has ceased, and with nothing to keep them large, white dwarfs contract until held up by the pressure of electrons forced so tightly together that they can get no closer. White dwarfs, the progeny of all stars with masses under ten times solar, are everywhere. The Sun will someday be among their host.

While DENEB and ALBIREO dominate the view of Cygnus, little 16 Cygni, a double star with an orbiting planet, steals at least some of the show. The class S star CHI CYGNI and the first parallax star, 61 CYGNI, show off as well. *(J. B. Kaler.)*

Residence:	CYGNUS
Other name:	HR 7503
Class:	G1.5 + G2.5 DWARFS
Visual magnitude:	5.96 AND 6.20
Distance:	70 LIGHT YEARS
Absolute visual magnitude:	4.30 AND 4.54
Significance:	THE FIRST DOUBLE STAR KNOWN TO HAVE AN ORBITING PLANET.

16 CYGNI

Imagine a double Sun. Not just a generic pair of stars, but a duo of solar clones in the sky, one setting right after the other, one's reddened light playing off the background of the other's colorful sunset. 16 Cygni comes close to that scenario, but not quite close enough, as Nature set these stars at least 850 AU apart. To get a double sunset, and for the pair to act gravitationally as one star, a planet would have to orbit in frigid cold thousands of AU away, surely precluding any life. Even then, the "Suns" would appear as no more than brilliant stars. From what we know about planet formation (which requires dusty disks around new **T TAURI** stars), such a planet would be impossible anyway. A single disk might form around a close double, but surely not around one whose members are that widely separated. The components of 16 Cygni, however, are so far apart that separate disks around each seems plausible. The result would be not just a double Sun, but a double planetary system.

16 Cygni appears as a mid-fifth magnitude star in northwest Cygnus. It is easily split with a small telescope to reveal two sixth magnitude solar twins, which, at a distance of 70 light years, are 39 seconds of arc apart. The brighter component, at 5785 K, is only 5 K warmer than our Sun and 60 percent more luminous; the fainter, though at 5660 K (20 K cooler than the solar surface), is also a bit (30 percent) brighter than the Sun, implying that each contains just slightly more than one solar mass. Though no orbit has been determined, it is clear that the stars are a pair, since they are at about the same distance and move together through space. The true separation of the stars cannot be known, because we cannot "see" the system in three dimensions, so it is impossible to tell which one is farther away from Earth. If they are at exactly the same distance (and therefore really 850 AU apart), it would take them 17,000 years to go around each other, making observation of orbital motion impossible.

Slight Doppler wobbles in the spectrum of the fainter star, 16 Cygni-B (of the sort seen for the original planet-holding star, **51 PEGASI**), reveal a planet, the only one known to belong to a double star. Since we do not know the tilt of the orbit, gravitational theory allows only a determination of a minimum mass of 1.5 times that of Jupiter. Unlike the close-in planet orbiting 51 Peg, this one behaves more (in our prejudiced opinion) "normally," orbiting 16 Cygni-B at a distance that averages 1.7 AU (12 percent larger than the Martian orbit) with a period of 804 days, or 2.2 years. The orbit is eccentric, however, and the planet comes as close as 0.56 AU to 16 Cygni-B (77 percent Venus's distance from the Sun), and then goes as far away as 2.8 AU, a bit over half Jupiter's distance.

Where there is smoke there is fire, and where a Jovian-type planet is seen, we wonder if perhaps an "Earth" might exist. Sadly, though, none-such can yet be detected. And while 16 Cygni-A so far seems barren, there is no reason not to expect it to have its own family. Imagine the view. From 16 Cygni-B's imagined "Earth," 16 Cygni-A would shine with twice the brilliance of our full Moon. The "16 Cygnians," seeing another "Sun" nearby, might, once their own system was explored, have a whole new planetary system they could visit, unlike Earth-dwellers who can only gaze longingly toward **ALPHA CENTAURI** and beyond. On the other hand, we should consider ourselves lucky because the presence of another star in the system, in addition to the looping orbit of a "Jupiter," might well destabilize the orbit of any "Earth" and make life impossible.

61 Cygni (see #86), the first parallax star (as well as a double star), sails through space, its image caught in 1916 (left) and 1948 (right). The parallax is only a third the size of the apparent stellar image, which because of the telescope optics and atmospheric turbulence appears as a disk. *(Lowell Observatory photographs.)*

Residence:	CYGNUS
Other name:	HR 8085 AND 8086
Class:	K5 + K7 DWARFS
Visual magnitude:	5.21 AND 6.03
Distance:	11.4 LIGHT YEARS
Absolute visual magnitude:	7.49 AND 8.31
Significance:	THE FIRST STAR TO HAVE HAD ITS DISTANCE MEASURED.

61 CYGNI

What are the stars? The conjectures go back to unrecorded time. Ponder the ancient argument about the nature of the then-visible Universe. Does the Sun go about the Earth, or the Earth about the Sun? We all know that it is the Earth that moves. Proving it, however, is not simple. It is easy, in fact, to show that it does not move. If you orbit a chair, the pictures on the walls swing back and forth. If Earth orbits the Sun, the stars should change their positions. To the eye, they do not. Ergo, we are at the center of the Universe.

Enter Copernicus. As we watch the planets, they swing back and forth, mostly moving to the east through the Zodiac. But as Mars approaches opposition to the Sun, it reverses direction and moves backwards, to the west. The Copernican explanation is that the Earth moves in solar orbit and that, as it passes between Mars and the Sun, the god of war just seems to reverse his march. If so, if Earth moves and the stars remain stationary, they must be terribly far away. They must truly be "Suns."

And if so, their apparent annual reflexive motions, their parallaxes, should—with enough telescopic power—be visible. With the improvement of the telescope, measures of stellar position became ever-more refined. Nevertheless, parallax remained elusive; the stars were farther than anyone imagined. Along the way, based on his observations of **GAMMA DRACONIS**, James Bradley discovered the aberration of starlight in 1728 (the 22-second-of-arc shifts in stellar positions caused by the orbital velocity of the Earth relative to the speed of light), which finally proved that Copernicus had it right. Twenty years later, Bradley discovered the 17-second-of-arc shift of nutation (a secondary wobble in the precessional wobble of the Earth's axis). Even with these shifts accounted for, still no parallaxes were visible. They had to wait nearly another century.

For Friedrich Bessel. He realized the best route to success was to observe the position of a nearby star. Flamsteed's 61 Cygni looked promising. Its high proper motion—angular movement across the sky—of over 5 seconds of arc per year, not far below that of **BARNARD'S STAR**, suggested closeness and made it a fine target. Sure enough, in 1838, Bessell announced a parallax of two-thirds of a second of arc (another proof of orbital motion). The formal parallax, half the shift, is now known to be 0.29 seconds of arc, from which we find a distance of 721,000 AU, or 11.4 light years. At last we could truly know the stars.

61 Cygni is a double whose members are the coolest and least-luminous dwarf stars visible to the naked eye (blended into one). The brighter K5 and fainter K7 components have respective temperatures of 4450 and 4120 K, luminosities of 0.145 and 0.091 solar, and masses of 0.6 and 0.5 solar. We have seen enough of the orbit to know that the pair takes 653 years to make a full turn at an average separation of 85 AU (over twice Pluto's distance from the Sun), the distance varying between 51 and 119 AU. The high proper motion, though fortuitously indicating closeness, is also in part the result of true high velocity, the stars—probably visitors from the Galaxy's halo—clipping along at 108 kilometers per second relative to the Sun. Consistently, the metal content is only about half solar. Moreover, both exhibit solar-like magnetic ("starspot") cycles. Each takes about 35 days to rotate. The brighter has a cycle of 7 years, the fainter a very Sun-like 11 years. Each is also classified as a "flare star" capable of sudden brightening caused by the release of magnetic energy. Bessel could not have picked better.

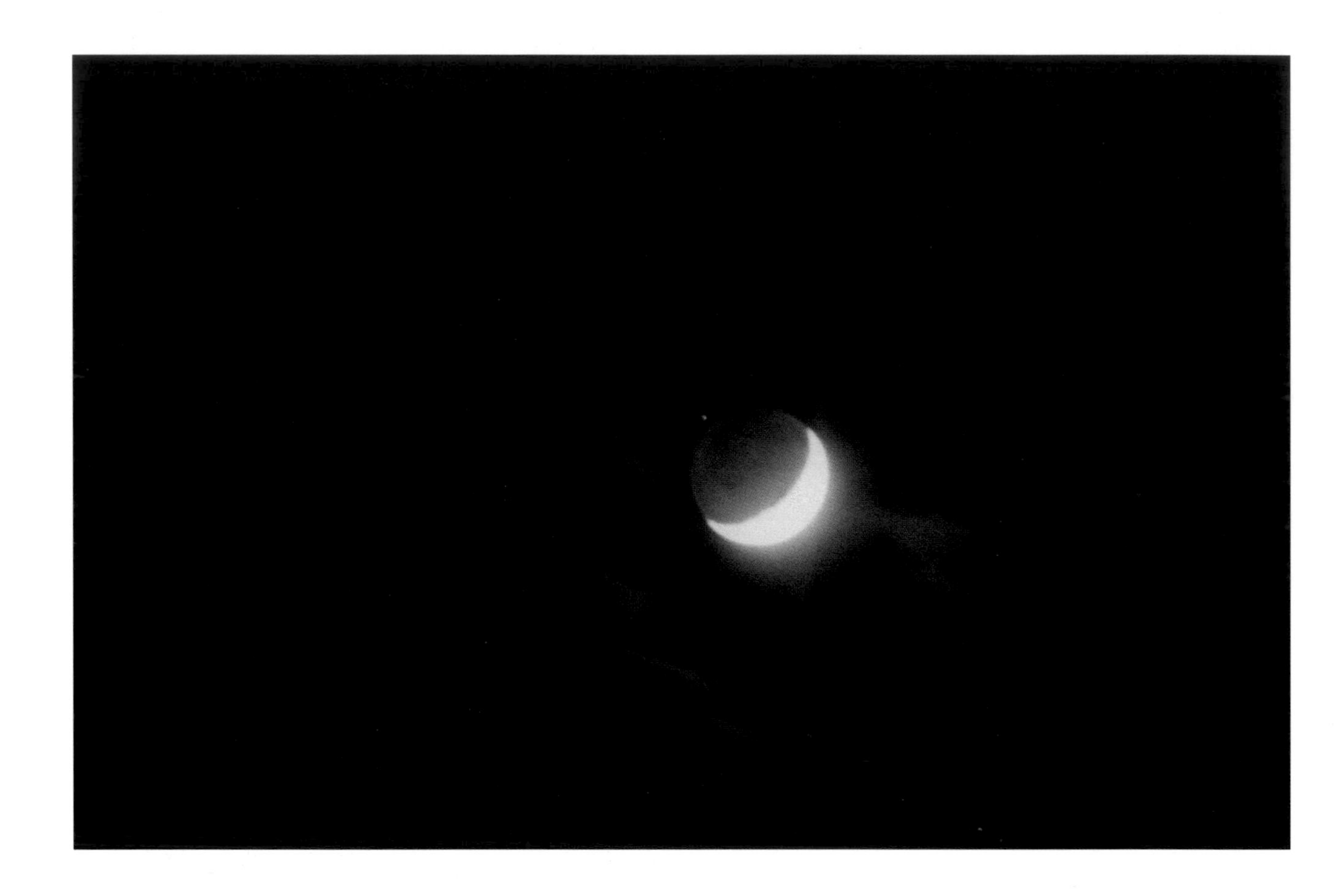

Stars near the ecliptic are commonly occulted by the Moon. Here, Eta Geminorum stands in for Spica, as the nighttime side of the crescent Moon, lit with earthlight, prepares to pass over it. *(J. B. Kaler.)*

Residence:	VIRGO
Other name:	ALPHA VIRGINIS
Class:	B1 + B4 DWARFS
Visual magnitude:	0.98
Distance:	262 LIGHT YEARS
Absolute visual magnitude:	–3.55
Significance:	BRILLIANT DOUBLE STAR CLOSE TO THE ECLIPTIC.

SPICA

Sixteenth-brightest star in the sky, in the Zodiac barely behind Aldebaran and Antares, Virgo's Spica helps illuminate the otherwise drab Northern Hemisphere's spring sky and serves as the southern anchor of the Great Diamond that connects the star to **ARCTURUS**, **COR CAROLI**, and Leo's Regulus. Among stars of first magnitude, Spica is second-closest to the ecliptic, following only Regulus, with the ecliptic stretched between them and the autumnal equinox lying right in the middle. As an "ecliptic star," Spica is often covered, or occulted, by the Moon. It is so bright that even when juxtaposed to the lunar disk, it remains readily visible. To find it, beginners follow the curve of the handle of the Big Dipper through Arcturus and then on to Spica. From Latin out of Greek, the word refers to a "sheaf of wheat" and is a reference to its ancient power as a fertility symbol, as the Sun at one time passed by it during the time of harvest. With an apparent magnitude of 0.98, Spica comes closer than any other star to defining "first magnitude." Of these brighter stars, only **ACRUX** and Beta Centauri exceed it in temperature and "blueness," though just barely. So admirable is Spica that it has two pointers, the Dipper's handle and the top two stars of Corvus, the Crow, which lies just to the west.

Spica is a multiple star that features a magnificent central double. Though the pair has been resolved via lunar occultations, it reveals itself best through the spectrograph, which shows absorption lines Doppler-shifting back and forth as the stars orbit each other. The period of revolution is but 4.01 days, showing the companions to be in tight proximity. As the two revolve, they also just barely eclipse, causing a dip in the brightness by just under one tenth of a magnitude.

Observation of the spectrum, of the eclipse, and of the actual orbiting stars through interferometry (the way in which the light waves from the stars interfere with each other) tell us that the pair orbits only 0.122 AU, or 26 solar radii, apart. The brighter and hotter star (22,400 K) has a luminosity 13,000 times that of the Sun, a mass 9 times solar, and a radius 8 solar. The companion, less than 10 percent as luminous, has about half that mass and radius. Complicating the picture, the larger star is also a slightly variable **BETA CANIS MAJORIS** star, jittering by a few hundredths of a magnitude over a 4.17-hour period.

The separation between the stellar centers is but double the summed stellar radii. The stars are so close that they tidally affect each other and are slightly out-of-round. The larger is beginning to evolve and is a good way through its hydrogen-fusing lifetime. Left alone, each star would become a massive white dwarf, the bigger of the two near the white-dwarf limit of 1.4 solar masses. Their evolution, however, will be influenced by duplicity. When the larger component begins to expand as a giant star, it will encroach upon its companion, and the two will begin exchanging mass. The larger star will eventually encompass the other, drawing it closer. Perhaps they will merge into one. Or perhaps reverse mass exchange from the dwarf left behind could create a nova (like **DQ HERCULIS**), even a supernova (like **TYCHO'S STAR**), as the white dwarf is pushed over the mass-limit edge. A front-row seat will be had by other companions, B5 and B7 stars at distances of 4 and 40 AU from the pair, and a 12th magnitude K dwarf over 10,000 AU away (if it is a member at all). Think of all that the next time the Moon makes a pass at Virgo's luminary.

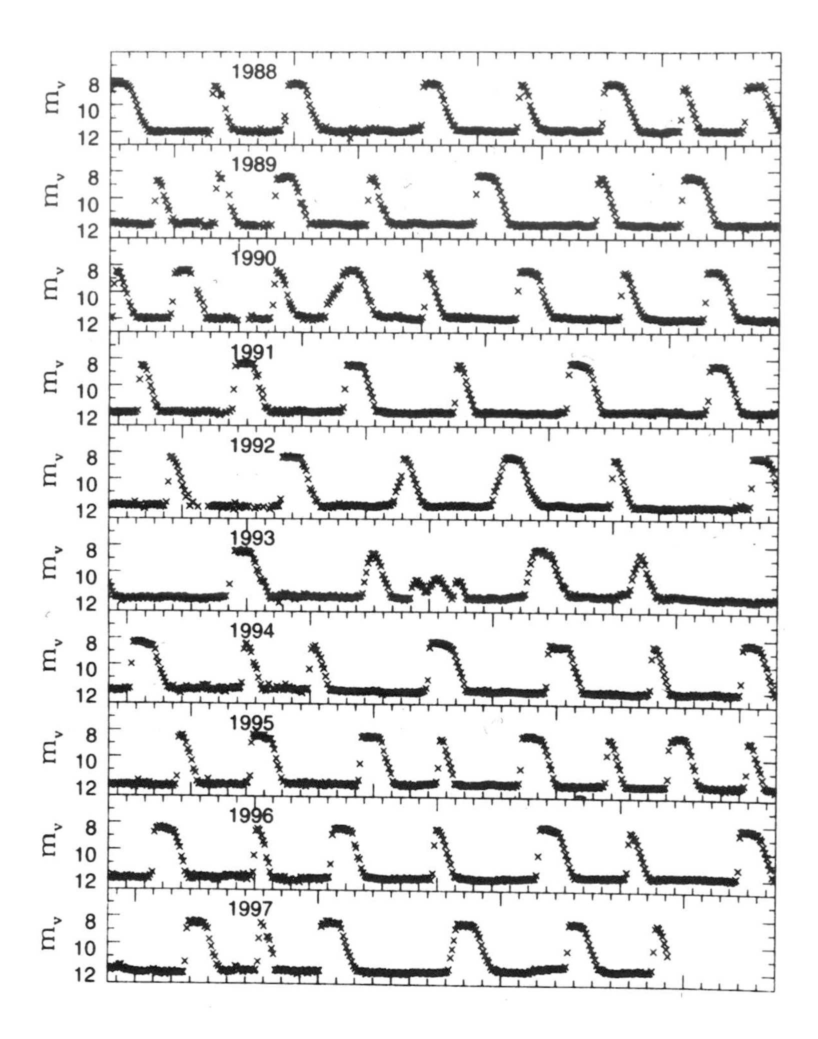

The dwarf nova SS Cygni entertains, popping off several times per year. *(Courtesy of the AAVSO. Cannizzo, J. K. and J. A. Mattei.* Astrophysical Journal *505 [1998]: 334.)*

Residence:	CYGNUS
Other name:	HD 206697
Class:	K4 DWARF + WHITE DWARF
Visual magnitude:	12 TO 8.2
Distance:	520 LIGHT YEARS
Absolute visual magnitude:	NOT APPLICABLE
Significance:	PROMINENT AND PROTOTYPICAL DWARF NOVA.

SS CYGNI

Next to bomb-like novae (**DQ HERCULIS**), or even recurrent novae (**RS OPHIUCHI**), dwarf novae like SS Cygni and U Geminorum resemble strings of Chinese firecrackers. Though the phenomena appear similar—SS's sudden outbursts just scaled down—their physical origins are entirely different. U Geminorum was the first of the breed to be found, SS Cygni the second. By virtue of its northern position, however, SS Cygni is by far the better-observed, as both professional and amateur astronomers have kept a near-continuous record of its behavior for over a century. About every 50 days, but impossible to predict, SS Cygni suddenly jumps from magnitude 12 to 8. The star may experience either a fast outburst, which takes place over a couple of days followed by a few days of "hang time," or a slower rise with longer staying power. The two types roughly alternate. The outbursts of the dwarf are seen all along the spectrum, even into the X-ray and radio.

Like real novae, dwarf novae are short-period binaries that consist of a white dwarf mutually revolving around a low-mass ordinary dwarf. The spectrum of SS Cyg and other U Gem stars display Doppler-shifting emission lines coupled to the white dwarfs from which, with other assumptions, we can infer system characteristics. SS Cyg consists of a high-mass (1.1 solar) white dwarf and a 0.7-solar-mass K5 ordinary dwarf. Separated only by the diameter of the Sun, they orbit in just 6.6 hours. They are so close together that the ordinary dwarf is tidally pulled into a teardrop shape until it meets the surface, at which the combined gravity of the two stars is effectively zero. From the point of the teardrop, which faces the white dwarf, matter flows into an accretion disk (which contains about 0.1 solar masses), from which it drops to the white dwarf, much as it does to create an ordinary nova like DQ Herculis. The stream from the K star hits the accretion disk with a glancing blow and creates a hot-spot. The glow that we see from the pair—even in quiescence—is not so much from either of the stars, but from the disk.

The sudden eruptions are caused not by any thermonuclear activity, but by an instability in the accretion disk that causes sudden brightening (visually by some ten solar luminosities). The idea is that if the mass-transfer rate is low enough, the temperature of the disk can be sufficiently low to allow the outer part of the dominating hydrogen in the disk to be neutral and opaque. The eruptions take place when the temperature of the gas climbs above the critical limit at which ionization occurs. As the outer part of the disk neutralizes again, the "star" dims. There are several variations on this theme: SU Ursae Majoris and its kind have "super-outbursts" that pop after ten or so ordinary outbursts, while the Z Camelopardalis stars have nearly continuous popping behavior, and, instead of dropping back to normal, can occasionally hang up at the halfway mark, sometimes for over one year.

If dwarf novae feed white dwarfs, then they should also produce thermonuclear novae. None has ever been seen among the set, which is not so surprising since we have not yet been observing them for the requisite 100,000 years. Some ex-novae, however, do display dwarf-nova behavior, lending credence to the idea. Those that do not either contain magnetic white dwarfs (the magnetism killing the accretion disks) or have too high a mass-transfer rate that keeps the disks hot.

For an evening's entertainment, U Gem and SS Cyg are easily seen with a backyard telescope. Maybe one will "nova" in front of your very eyes.

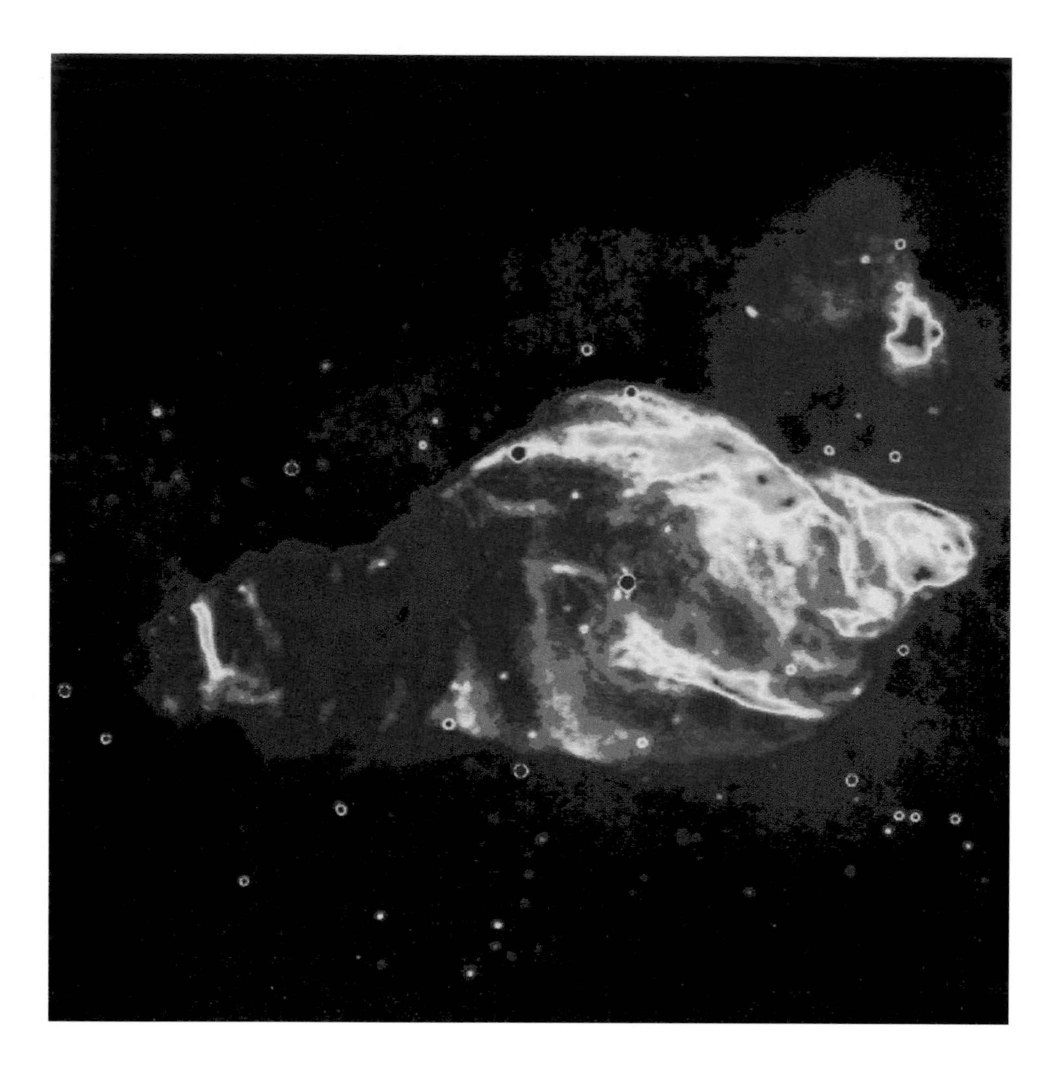

The huge supernova remnant W50, imaged here in radio radiation, encloses SS 433 and its speedy jets. *(Very Large Array, National Radio Astronomy Observatory, G. M. Dubner, M. Holdaway, W. M. Goss, and I. F. Mirabel, from a paper in the* Astronomical Journal *116 [1998]: 1842.)*

Residence:	AQUILA
Other name:	V 1343 AQUILAE
Class:	OB SUPERGIANT + NEUTRON STAR
Visual magnitude:	14.2
Distance:	10,000 LIGHT YEARS
Absolute visual magnitude:	–6
Significance:	NEUTRON STAR WITH WOBBLING HIGH-SPEED EJECTA.

SS 433

Though to us they may seem high, most astronomical speeds are low, with stars moving at a few tens of kilometers per second. By contrast, supernova ejecta, which scream outward at 10,000 kilometers per second, seem extraordinary. Yet even these look pale beside what is coming from the emission-line star SS ("Stephenson and Sanduleak") 433. While its spectrum shows normal emission lines of hydrogen and helium, it also displays dual sets of hydrogen and helium emissions that alternately shift back and forth in wavelength over a huge stretch of spectrum, crossing each other with a 164-day period. They show gas both receding from us at an amazing 50,000 kilometers per second and approaching at 30,000, with an average implying a recession close to 10,000 kilometers per second. The stationary emissions, however, show SS 433 to be moving away at only 70 kilometers per second. The key to understanding the star lies in its environment. Ten thousand light years away, it is in the middle of a huge gaseous supernova remnant, W50. Some kind of collapsed object with a great gravitational field could be the culprit, one that ejects powerful flows of matter.

The "stationary" emissions are not quite so. They Doppler-shift slightly over a 13-day period, revealing that the visible star mutually orbits an invisible companion, probably a neutron star at a separation of no more than a couple tenths of an AU. Matter flows from the visible star, an O- or B-type supergiant, into a rotating "accretion disk" and then down to the neutron star itself. Instead of being accreted, however, some of the matter is whirled outward at enormous speeds to create the jets that radiate the madly shifting emission lines.

The axis of the Earth, tilted by 23.5 degrees relative to the orbital axis, wobbles, or "precesses," with a 26,000-year period. The cause is the gravitational action of the Moon and the Sun on the Earth's equatorial bulge. The axis of the accretion disk surrounding SS 433's neutron star must be tilted as well. The gravity of the normal star causes the accretion disk to precess like the Earth, but over a much shorter, 164-day period. As a result, the high-speed jets sway like a berserk lawn sprinkler, causing the wildly moving spectral emissions. When the geometry of the system is accounted for, the jets must emerge at 26 percent the speed of light. The huge average radial velocity of the emissions is an effect of Einstein's theory of relativity and the great speed of the jet material as it pours out of the gravitational depths of the collapsed object. SS 433 thus provides another proof of this famous theory (as if any further evidence is actually needed).

Activity of this magnitude has had a profound effect on the star's surroundings. With radio telescopes, we see corkscrewing gaseous streams a few seconds of arc across coming out of the star. These are aligned with a bar of X-ray emission nearly 1 degree long, the jets pounding the local gaseous medium to a distance of more than 100 light years from the neutron star. The corkscrews and X-ray emissions are also aligned with the major axis of the supernova remnant and may have helped shape the structure. The double star and W50 (nearly 200 light years across) are in the act of shoveling 30,000 solar masses of surrounding interstellar gas. Objects like **GRS 1915+105**, which are suspected to contain black holes, show higher speeds; but none dances like SS 433.

The Hubble Space Telescope beautifully captures the expanding remnant of Supernova 1987A within the glowing extended nebulosity of the Large Magellanic Cloud. The triple-ring structure, some five light years across, is caused by light from the supernova ionizing and illuminating matter lost by the star before it blew up. *(Hubble Heritage Team [AURA/STScI/NASA].)*

Residence:	DORADUS
Other name:	SANDULEAK –69°202
Class:	TYPE II SUPERNOVA
Visual magnitude:	2.9 AT PEAK
Distance:	163,000 LIGHT YEARS
Absolute visual magnitude:	–15.6
Significance:	BRIGHTEST SUPERNOVA IN 400 YEARS.

SUPERNOVA 1987A

High atop a Chilean mountain in February of 1987, Ian Shelton was developing the last photographs he had taken of one of our companion galaxies, the Large Magellanic Cloud, when he noted something was "wrong." There was a brilliant interloping star where none had been before. A quick peek outside confirmed its existence. Shelton was looking at the first naked-eye supernova since **KEPLER'S STAR** of 1604. And he was not alone. Robert McNaught in Australia had caught it on the way up the day before. A dry spell of nearly four centuries had been broken.

By May, Supernova 1987A peaked in visual brightness at third magnitude. Over 150,000 light years away, the exploder had a maximum visual luminosity 150 million times that of the Sun! Rather than studying dim supernovae in distant galaxies, we could finally observe a bright exploding star in great detail with modern equipment. Examination of the field of stars prior to the explosion quickly revealed the source: a blue (class B3) supergiant catalogued as "Sanduleak –69°202," a former 20-solar-mass star that is no longer there.

Supernova 1987A was a classic "iron-core collapse" Type II supernova very much like the one that created the **CRAB NEBULA** in 1054. It provided powerful proof that our theories about Type II supernovae are correct. Over millions of years, a high-mass star will build progressively heavier elements in its core as it supports itself by thermonuclear fusion. When we arrive at iron, the core contracts with devastating consequences. Well within a second, it drops from the size of Earth to a ball only a few tens of kilometers across. While the iron breaks down into its components and then into neutrons, vast numbers of near-massless, speed-of-light neutrinos are released. (Neutrinos, a natural by-product of fusion reactions, continuously flood from the Sun.) The shock wave caused by the collapse blows the rest of the star apart. It takes a day or more for the shock to reach the stellar surface and for the star to brighten visually, but the neutrinos scream out right away. Telescopes designed to monitor solar neutrinos produced by hydrogen fusion in the solar core caught a total of 11 of them from SN 1987A, about the number theoretically expected from such a distant event.

During these high-temperature explosions, supernovae produce vast amounts of heavy elements, including a tenth or so of a solar mass of radioactive nickel, which decays to radioactive cobalt and then into iron. All of the iron in the Universe has come from Type II and Type Ia (**TYCHO'S STAR**) explosions. The heat generated by these and slower decays powers the optical light curve—brightness vs. time—of the explosion. Proof comes from gamma rays directly detected from the dying cobalt.

The flash from the 1987 explosion illuminated matter that had been lost through powerful winds thousands of years before, creating a set of graceful funnel-like rings up to a few light years across. A dozen years after, the blast wave began to hit this extended envelope, lighting it further. At the core is the debris of the explosion itself, expanding at 2500 kilometers per second (marked down from an original 30,000). At the center we expect to find a neutron star, though none is seen. Perhaps there is a black hole instead. Or perhaps there is nothing at all.

T Tauri, prototype of all young stars, is surrounded by a complex of small nebulae, and is the origin of a giant flow that from end to end is five light years long. *(B. Reipurth, J. Bally, and D. Devine, CTIO Curtis Schmidt Telescope, image courtesy of John Bally.)*

Residence:	TAURUS
Other name:	HD 284419
Class:	T TAURI PROTOSTAR
Visual magnitude:	9.6 VARIABLE
Distance:	460 LIGHT YEARS
Absolute visual magnitude:	2.0 VARIABLE
Significance:	PROTOTYPE FOR NEW STARS.

T TAURI

In 1945, Alfred Joy of the Mount Wilson Observatory established the criteria for a new class of variable stars, the T Tauri stars: "(1) irregular light variations of about 3 magnitudes, (2) spectral type F5–G5 with emission lines resembling the solar chromosphere, (3) low luminosity, and (4) association with dark or bright nebulosity."[1] The standards still hold. As so often happens, however, the prototype is not very exemplary of the rest of the crowd. Indeed, the founding member of the group is unique.

Associated with the dark, dusty molecular clouds of its formation and like others of its kind, T Tauri is youthful, a brand-new star. It is approaching main-sequence stability, when it will quietly fuse its core hydrogen into helium, and is a successor to hidden objects such as **L 1551 IRS 5**. At first, it looks something like our Sun, with a G-type spectrum. However, it is much more luminous than the Sun, a clear giveaway of newness. It is still in a state of erratic mass accretion and gravitational contraction, causing its brightness to flop around by over three magnitudes over the course of weeks or months; even the spectral class changes from G4 to G8. The prominent emission lines include lithium, an element so easily destroyed by high temperatures within stars (as it circulates downward by convection) that its presence again tells of youth. T Tauri also radiates strongly in the ultraviolet. This high-energy radiation is produced by a circumstellar disk, from which the star accretes mass and whose cooler outer dust glows powerfully in the infrared. The star is associated with two nebulosities, Hind's Variable Nebula (a reflection nebula) half a minute of arc to the west, and the much closer (and also variable) Burnham's Nebula.

T Tauri's real novelty involves its multiplicity. Imaging at infrared wavelengths reveals a second source about 0.9 seconds (125 AU) to the south of the optically visible star. When measured across the entire spectrum, the second star (T Tauri South) is more luminous than the one we see (now known as T Tauri North). Yet it remains utterly invisible at optical wavelengths. The distribution of infrared energy suggests that the object is dimmed 35 magnitudes to the eye by intervening dust. How T Tauri North can be so bright and visible and T Tauri South so invisible is a mystery. As a result, little is known about T Tauri South. It may be a protostar at an earlier stage of development, it may be hidden behind T Tauri North's disk, or maybe something else is at work. We do know, however, that it is itself double, two new stars only 7 AU apart, showing us a triple star in formation.

In the 1950s, George Herbig and Guillermo Haro discovered odd nebulae that seemed to have no sources of illumination. Herbig-Haro objects come in pairs with a T Tauri (or similar youthful object) between them. These "HH objects" are regions of shocked gas shoveled up by massive bipolar flows of molecular gas that emerge in opposite directions from the disks of new stars. T Tauri is associated with a "giant flow" over half a degree—5 light years!—long. The inner portion, which includes Burnham's Nebula, is a mess, probably as a result of flows from all the component stars. Rather than coming from optical T Tauri North, however, the giant flow is coming from T Tauri South.

Someday, the dust will clear. In the distant future, we will see an ordinary triple, two (probably class F) stars in tight orbit, with a third going around the pair. The one we now know as T Tauri is setting the stage for all the others of its kind.

[1] Joy, A.H. (1945). "T Tauri Variable Stars." In *Astrophysical Journal* 102: 168.

The famed Trapezium, led by Theta-1 Orionis C, illuminates the stunning Orion Nebula, seen here in infrared light. *(Very Large Telescope, courtesy of M. McCaughrean.)*

Residence:	ORION
Other name:	TRAPEZIUM
Class:	B1 + B0 + O6 + B0.5 DWARF QUADRUPLE
Visual magnitude:	6.7, 8.0, 5.1, 6.7: 4.7 COMBINED
Distance:	1400 LIGHT YEARS
Absolute visual magnitude:	–3.2, –1.9, –4.8, –3.2
Significance:	THE CENTRAL EXCITING STARS OF THE ORION NEBULA

THETA-1 ORIONIS

Centered in Orion's Sword 1400 light years away, the Orion Nebula, a classic diffuse nebula 20 light years across, is energized by a quartet of stars 22 seconds of arc (10,000 AU) wide that is collectively called Theta-1 Orionis, or the Trapezium. Though the nebula looks as if it surrounds these hot, blue O and B stars, most of it is a heated blister on the surface of a huge dark, dusty, molecule-filled cloud that lies behind much of the constellation and that is in a state of active star formation. The Trapezium was born from it only a few hundred thousand years ago and is now eating part of the cloud away. The stars' ultraviolet radiation is killing the cloud's molecules and ionizing the released atoms, which are mostly hydrogen. Recapture of electrons by protons and other nuclei cause the nebula to glow in the same manner as the planetary nebulae **NGC 6543** and **NGC 7027**.

The Trapezium stars are lettered from west to east A, B, C, and D. As a unit, they generate 250,000 solar luminosities and are at the pinnacle of one of the densest stars clusters in the Milky Way. The relation between class and apparent brightness is confused. Though B is notably the dimmest, it appears to be hotter than both A and D. The problem involves uncertainties in classification (A has been listed as hot as O7), correction for dimming by foreground dust, and the effects of orbiting companions.

The leader of the pack is magnificent fifth magnitude Theta-1 Orionis C. With a spectral class of O6, it radiates an awesome 210,000 solar luminosities, 85 percent of the foursome's total, from a 40,000-K surface. Since C is by far the hottest of the quartet, it generates 95 percent of the ultraviolet light that ionizes the nebula. The great luminosity of this 40-solar-mass star produces a 1000-kilometer-per-second wind that blows with a flow rate over 100,000 times that of the solar wind. There is some evidence that it can change its spectral class to as hot as O4. The star is so luminous that it is destroying the dusty disks around the young **T TAURI**-like stars nearby that would, if left alone, probably create planets.

A and D are nearly tied for number two in apparent brightness. A, also known as V ("variable") 1016 Orionis, was only discovered in 1973 to be an eclipsing double with a markedly long period of 65.4 days. Though the eclipse is nearly one magnitude deep, it lasts a mere 20 hours, making the event difficult to observe. The companion is thought to be a 2.5-solar-mass T Tauri star. Averaging only 1 AU apart, the pair can approach as close as 0.19 AU. Another companion lies 100 AU away, making A a triple star. D seems simplest, though it too seems to have a companion made known only by its effect on D's spectrum. That leaves the dimmest to the eye, Theta-1 Orionis B. Sixty AU away orbits a class B companion called B1, while B proper (also called BM Orionis) is yet another eclipser. Every 6.47 days, it dims by 0.8 magnitudes as a probable class G star moves in front of it. B1 is also binary, making B a quadruple star!

To survive, multiple stars must be hierarchical (doubles in orbit about doubles, for example), which the Trapezium distinctly is not. Two of the stars will eventually be accelerated by gravitational interactions and be kicked out (perhaps making runaway stars like **MU COLUMBAE**). The remaining pair will probably be separated by the effects of neighboring stars. After parting, C is destined to blow up as a supernova. The others, each around a dozen solar masses, may explode or make massive white dwarfs. But do not despair. The Orion Molecular Cloud probably has more "Trapeziums" in the waiting.

Thuban (see also #44) was the pole star at the time of the building of the Great Pyramid of Cheops. *(Gay Robins.)*

Residence:	DRACO
Other name:	ALPHA DRACONIS
Class:	A0 GIANT
Visual magnitude:	3.65
Distance:	310 LIGHT YEARS
Absolute visual magnitude:	–1.23
Significance:	THE ONCE AND FUTURE POLE STAR.

THUBAN

Shining at fourth magnitude between **MIZAR** and the bowl of the Little Dipper, Thuban is one of the faintest stars to have a famed proper name, the result not of physical glory but of position. Its location was probably why Bayer made it Draco's Alpha star, even though it ranks number seven in brightness. The Dragon's brighter stars are named rather in order from head to tail, but here is "Alpha" stuck in toward the tail's end, testifying to the star's importance. "Thuban" ("serpent" in Arabic) was erroneously derived from a fearful corruption of the original Arabic name for Rastaban (Beta Draconis)—and then applied to the wrong star.

Thuban's claim to stellar significance is that it is about as bright as a class A star can be and still live on the edge of being a hydrogen-fusing dwarf. From a distance of 310 light years, the star shines some 300 times more brilliantly than does the Sun, its white surface heated to around 9800 K. Though formally classed as a giant, this 3.4-solar-mass star has either just stopped fusing its core hydrogen into helium or will do so very shortly. It would be more accurately described as a subgiant. Whatever the details, it is about ready to make its rapid transition to a brilliant and swollen red giant. Though many white class A and B stars, especially those that spin as slowly as this one (its rotation period about that of the Sun), are chemically peculiar. Thuban is not, though, having a notably lower metal content than the Sun. Accompanying it in a 51-day orbit is a companion (known only through the spectrum) of unknown character. The two are probably no more than half an AU apart, which will do the little guy no good when the big one, our Thuban, begins to grow. As it encroaches on the companion, it may create an **ALGOL**-like mass-transfer binary. If the smaller star, transformed into a white dwarf, is dragged inward and encompassed by a common envelope generated by Thuban, the pair may eventually create a nova.

But back to Earth. Our Pole Stars are now **POLARIS** in the north and dim **SIGMA OCTANTIS** in the south. But it was not, and will not be, forever so. Because of the 26,000-year precession of the Earth's axis, the celestial poles move in large circles 23.5 degrees in radius around the fixed orbital (ecliptic) poles. Still moving a bit toward Polaris, the northern pole will next take a bead on Cepheus. Going backward, around the time of the Greek poet Homer, Kochab in Ursa Minor was a (rather poor) Pole Star. Among the best ever, however, was Thuban, which lay almost exactly at the pole in 2700 BC. It remained better than Kochab up to around 1900 BC and was the Pole Star during the time of the ancient Egyptian civilizations. It was such a wonderful polar marker that, for over 150 years, it was believed that the entry tunnel to the Great Pyramid of Cheops was deliberately pointed to it, allowing scholars to date the pyramid. The entrance was directed upward at 26.5 degrees from the horizontal (but not to the pole itself), and there were two eras when the star would have crossed the tunnel's view, once as Thuban approached the pole, once when it receded.

Alas, the brilliant idea does not work and gives incorrect dates. The tunnel is tilted according to how the blocks were laid on one another, and the closeness of Thuban to crossing its line of sight is a coincidence. None of that, however, takes away from what the Egyptians saw: Thuban near the pole, Polaris rounding it 26 degrees away. We will see it all again around the year 25300.

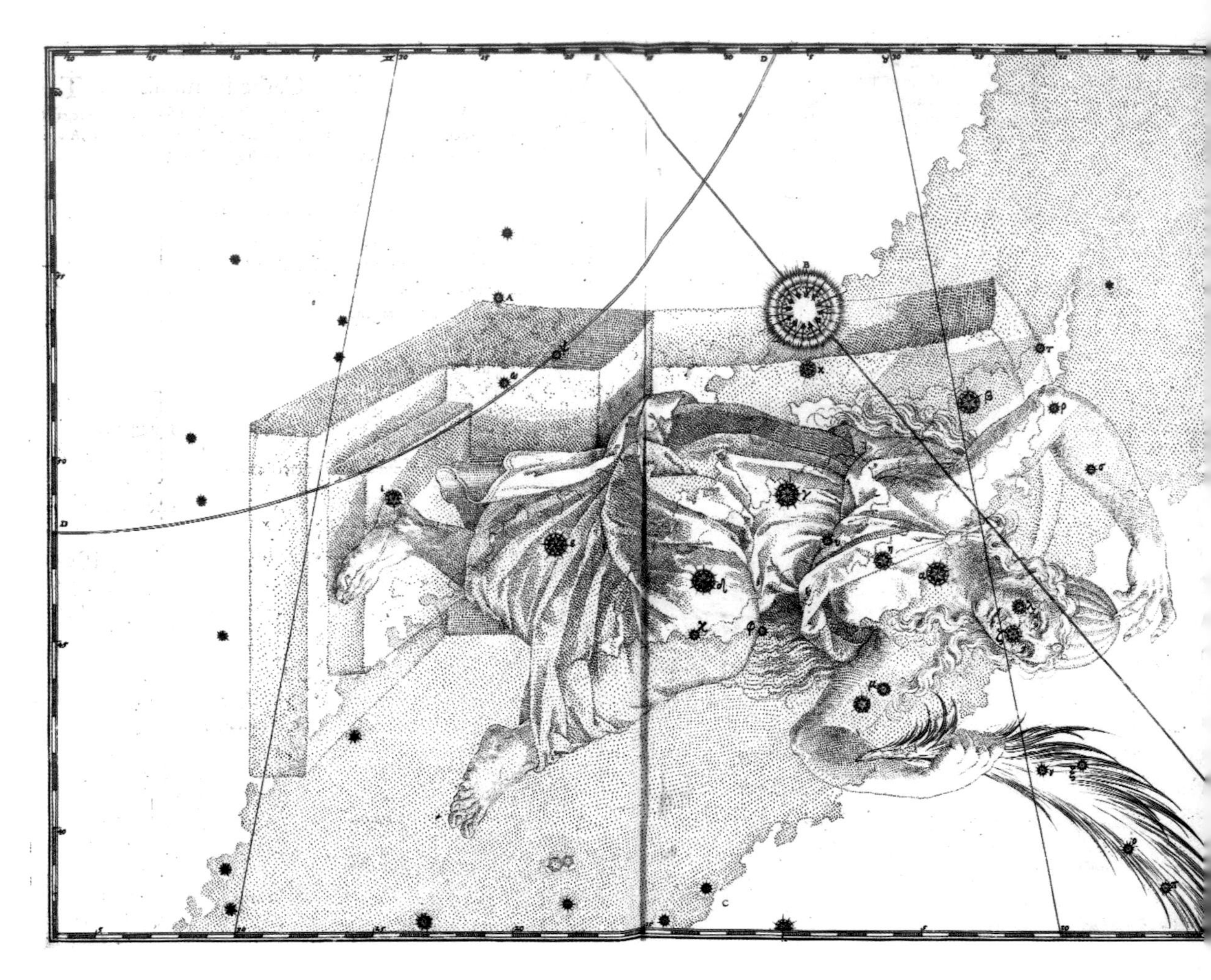

Tycho's Star (the large symbol above Cassiopeia's Chair) was so important that even though it had long-since faded by the time Johannes Bayer published his great Uranometria in 1603, it was still immortalized. (*Johannes Bayer. 1603.* Uranometria. *Courtesy Rare Book Room and Special Collections Library, University of Illinois.)*

Residence:	CASSIOPEIA
Other name:	SN 1571
Class:	TYPE IA SUPERNOVA
Visual magnitude:	–4 (MAXIMUM)
Distance:	11,500 LIGHT YEARS
Absolute visual magnitude:	–19.1 (MAXIMUM)
Significance:	THE SUPERNOVA THAT CHANGED THE WORLD.

YCHO'S STAR

Vovember 6, 1572, W. Schuler of Wittenberg
essed a brilliant glowing coal in Cassiopeia, one
outshone all the stars around it. Five days later—
lelay likely caused by the autumn clouds endemic
orthern Europe—Tycho Brahe discovered the "new
' the "nova stellarum," for himself, and the courses
tronomy and human thought changed forever.
used, in part, by the new star, Tycho founded an
rvatory where he made detailed measures of the
ions of ordinary stars and of the movements of
ets. These observations were later used by Johannes
er to establish the laws of planetary motion that, in
helped Newton to define the law of gravity. Less
t were Tycho's observations of the nova itself. For
r and a half, he kept detailed records of its bright-
relative to other celestial bodies.
Novae are not uncommon. One was even seen in
ndromeda Nebula, a massive spiral galaxy. Only
ı Edwin Hubble discovered that such "spiral
lae" were external systems did we realize that their
e—like the 1885 display were actually super-
e. The brilliance of Tycho's Star suggested that it
vas a supernova, as were **KEPLER'S STAR** of 1604
he Chinese Guest Star of 1054 that made the
B NEBULA. In 1945, Walter Baade of the Mount
on Observatory used Tycho's observations to
ıstruct the supernova's visual "light curve" (magni-
plotted against time). At its peak, it shone as a
ıd "Venus" near magnitude –4 and was visible in
aylight. Fading to Jupiter's magnitude, then to
of Cassiopeia's stars, it finally dropped below
d-eye visibility in March 1574.
Supernovae divide into two broad kinds. Type II
s derive from iron-core collapse in massive stars
and leave behind both gaseous remnants and (usually) neutron stars. The more powerful (and brighter) Type Ia are caused by white dwarfs in double stars that accrete too much matter from binary companions. Overflowing the white-dwarf mass limit, they blow up and annihilate themselves. Both kinds initiate powerful shock waves that sweep up the interstellar medium and radiate from the radio spectrum through the X-ray. The shape of its light curve suggested that Tycho's Star was "Ia." However, measures of distance and the resulting defining luminosity have been devilishly difficult. We can estimate distance from a variety of methods that include the rate of angular expansion and the amount of interstellar gas in front of the expanding remains (which produce spectral absorptions). We must, however, also account for the degree of absorption of the light by interstellar dust. Fortunately, Tycho and others even left behind observations of the supernova's color, which (since dust also reddens starlight) allowed the observed magnitudes to be corrected. The best estimate is nearly 12,000 light years, which yields an absolute visual magnitude of –19.1 (4 billion times the solar visual luminosity), within the range of expected values.

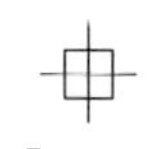

The death of a white dwarf over 400 years ago deposited a third of a solar mass of iron into the cosmos, helped give us the greatest of the pre-telescopic astronomical discoveries, and left behind a ring glowing with X-rays. Because of their luminosity and uniformity, Type Ia supernovae are used to probe the distant reaches of the Universe and to establish the properties of its expansion. Perhaps millions of years from now an astronomer in some distant galaxy will use Tycho's Star for the same purpose.

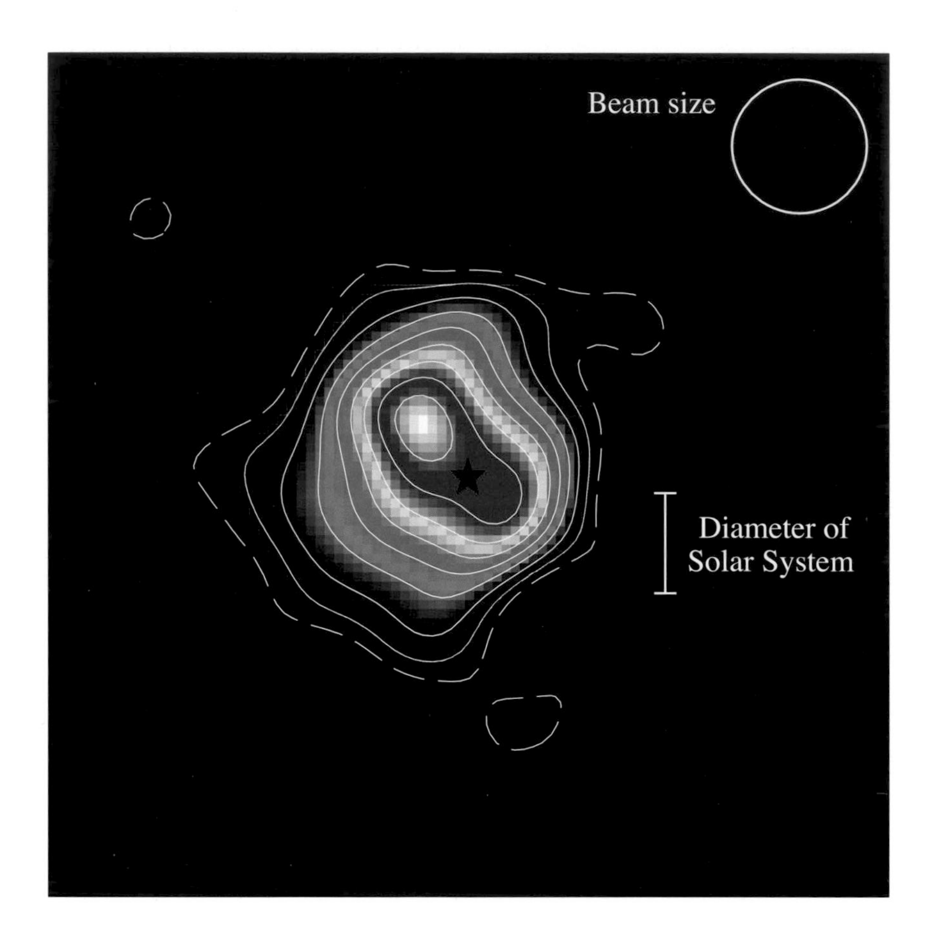

Vega (see also #35) is surrounded by a cloud of warm dust left over from the time of its formation. *(W. S. Holland et al., from an article in* Nature *392 [1998]: 788.)*

Residence:	LYRA
Other name:	ALPHA LYRAE
Class:	A0 DWARF
Visual magnitude:	0.03
Distance:	7.76 LIGHT YEARS
Absolute visual magnitude:	0.58
Significance:	A LUMINARY OF THE NORTHERN HEMISPHERE.

VEGA

The three brightest stars, **SIRIUS**, **CANOPUS**, and **ALPHA CENTAURI**, are all in the Southern Hemisphere. The Northern Hemisphere contains the next three, **ARCTURUS**,Vega, and **CAPELLA**. Though the southern trio spans a considerable range of apparent magnitude, the northern are separated by a mere 0.12, Vega in the middle, 0.08 magnitude fainter than Arcturus. While Capella defines the northern winter and Arcturus announces spring, Vega hangs nearly overhead in mid-northern latitudes, looking lazily down on summer's heat. The luminary of Lyra, Orpheus's celestial harp, "Vega" is a corruption of an Arabic phrase that refers to a "swooping eagle," appropriate for such a high-flying star.

Vega is the ultimate standard star. Defining "white," its photographic magnitude (appropriate to blue light) and visual magnitude (appropriate to yellow) are identical, helping relate color to temperature (Vega's pinned at 9600 K). The star has long served as a magnitude standard because its apparent and photographic magnitudes are very close to 0.0. All other stars are ultimately compared with it, though by way of ever fainter cross-matched standards, as Vega is so bright that it would destroy modern detectors at large telescopes. The distribution of energy across its spectrum also provides a standard for stellar spectroscopy.

Like five of the top six (Canopus a distinct exception), Vega can dominate the sky because it is so nearby, only 25.3 light years away. A hydrogen-fusing dwarf of 1.5 solar masses that radiates 54 solar luminosities, Vega's age is calculated at 385 million years, 10 percent the solar age and (since higher-mass stars age faster) 65 percent of the way to the exhaustion of the fuel in its core.

Though at this point seemingly ordinary, Vega harbors some secrets. Many white stars of classes A, B, and F are chemically peculiar with huge over-abundances of certain elements, underabundances of others. Vega, however, is a "low-metal" star. Its general metal content is only 25 percent of the Sun's, much too low for a "local" star and more like those of the Galaxy's halo, to which it does not belong. No explanation is available. A seemingly slow rotator (as are those of the chemically peculiar persuasion), it seems to have its rotation pole pointed at the Earth.

The biggest Vegan surprise was a disk of cool dust discovered in 1984 by the Infrared Astronomical Satellite. Vega's disk, chilled to around 75 K, extends up to 70 AU from the star. Our Solar System has such a disk made of the debris of the formation of the planets, of broken asteroids, of comet dust. We have chased the solar disk into the "Kuiper Belt" of comets to a very similar 55 AU. Vega's disk also has a hole in it, that is, the middle portion has fewer dust particles. Could Vega have planets as well? None has been seen, but then, if the plane of its system is perpendicular to the line of sight, there would be no back and forth motion of the star (as is the case for cooler **51 PEGASI**).

Alas, even if there are planets, life could barely get a foothold. At an age of 385 million years, the Earth was still being bombarded by the debris of the early Solar System (as testified by lunar cratering). By the time life arose, Vega would be dying.

9

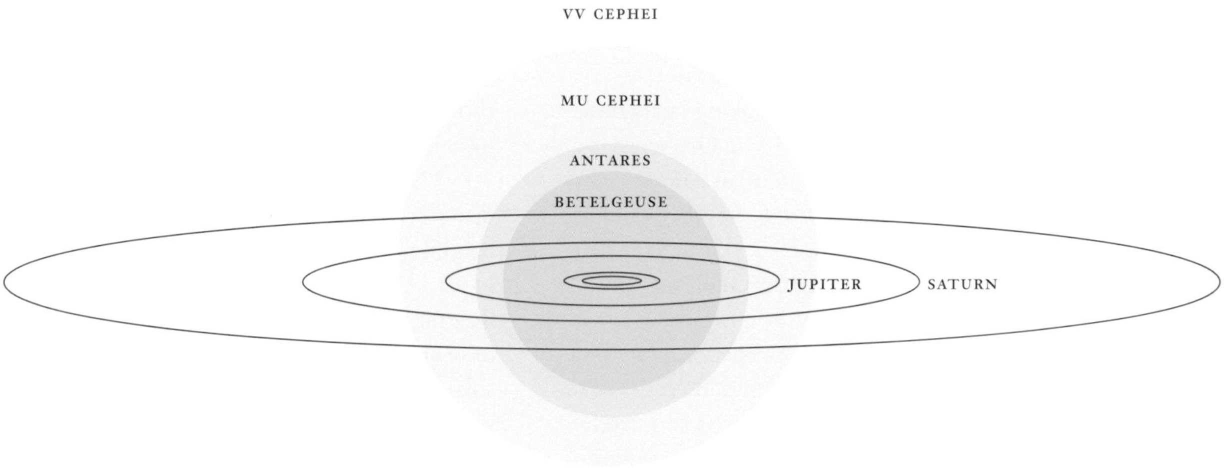

VV CEPHEI
MU CEPHEI
ANTARES
BETELGEUSE
JUPITER
SATURN

Approaching the size of Saturn's orbit, VV Cephei is the largest star known, topping even MU CEPHEI, and more famed ANTARES and BETEGEUSE. *(Adapted from J. B. Kaler. 2001.* Extreme Stars. *Cambridge: Cambridge University Press.)*

Residence:	CEPHEUS
Other name:	HR 8383
Class:	M2 SUPERGIANT + B8(?) DWARF
Visual magnitude:	4.91
Distance:	2000 LIGHT YEARS
Absolute visual magnitude:	–6.75
Significance:	LONG-PERIOD ECLIPSER AND THE LARGEST STAR KNOWN.

VV CEPHEI

Not very impressive as a constellation, Cepheus's kingly stature is secured through a pair of his stars, two of the gargantua of the Galaxy: the red supergiants **MU CEPHEI** and VV Cephei. Respectively fourth and fifth magnitude, both shine from within the Cepheus OB2 association, a vast grouping of O and B stars about 2000 light years away. Itself too far for accurate parallax measure, VV's distance is estimated from the average distances of the stars within the grouping. Accounting for two magnitudes of absorption by interstellar dust, VV glows with an absolute visual magnitude of –6.8.

Unlike Mu Cephei, VV is not content to be alone and is accompanied through life by a hot dwarf whose class (somewhere between O and A) is highly uncertain. At visual wavelengths, the red supergiant dominates, supplying about 80 percent of the combined light. With that small correction and a much larger one for infrared radiation from its cool 3300-K surface, VV has a total luminosity 315,000 times that of the Sun. From these properties, we derive a radius of 8.0 AU, 85 percent the size of the orbit of Saturn, which makes VV the largest star known.

Orbital analysis provides confirmation. Every 20.2 years, the B dwarf (or whatever it is) hides completely behind the M supergiant. From Earth's perspective, the little one takes 250 days or so to disappear, just barely eclipsing behind the top of the larger star. The B star then hides for about one year, taking another leisurely year to reappear. Given the visual dominance of the supergiant, the event is only a slight dip to the naked eye. In the part of the ultraviolet spectrum that sneaks through the Earth's atmosphere, where the hot B star dominates, the combined light drops by an impressive 2.5 magnitudes. At yet shorter wavelengths, those only accessible to space observatories, VV effectively disappears, "lights out" for a whole year.

The light variation combined with stellar velocities yields the nature of the rather elliptical orbit. Averaging 25 AU apart, the stars come as close as 16 AU. They are so near each other that tides cause mass to flow from the supergiant toward the dwarf, where it circulates in a disk over 2 AU across before crashing onto the dwarf's surface. At close approach, tides may even cause the supergiant to fill its "zero-gravity surface" (where the effective gravity of one star balances that of the other), strongly enhancing the flow. So much matter is lost that the orbital period changes: the 1997 eclipse arrived 68 days late! Radiation from the streaming matter confuses the stellar observations, making analysis difficult.

The duration of the eclipse gives an approximate radius for the supergiant of (drum roll here) 8.8 AU, bigger than that derived from the luminosity and temperature! Even a conservative analysis yields 7.4 AU. This grand supergiant is roughly half the size of the B star's orbit. So huge as to be unstable, the red star is separately classed as a semi-regular variable whose visual brightness changes over a two-month period by around 0.3 magnitudes, further complicating analysis of the orbit.

Both stars are estimated to contain around 20 solar masses. At one time, they were a pair of blue-white dwarfs. The more massive of the two evolved first and is now the red supergiant. It will surely explode, perhaps sending the B star flying away as a "runaway star," or making itself into the B star's neutron-star companion. Then it will be the B star's turn to explode, and a double neutron star may be created. But that is in the future. For now, VV Cephei, ruling the Galaxy, is the power behind King Cepheus's throne.

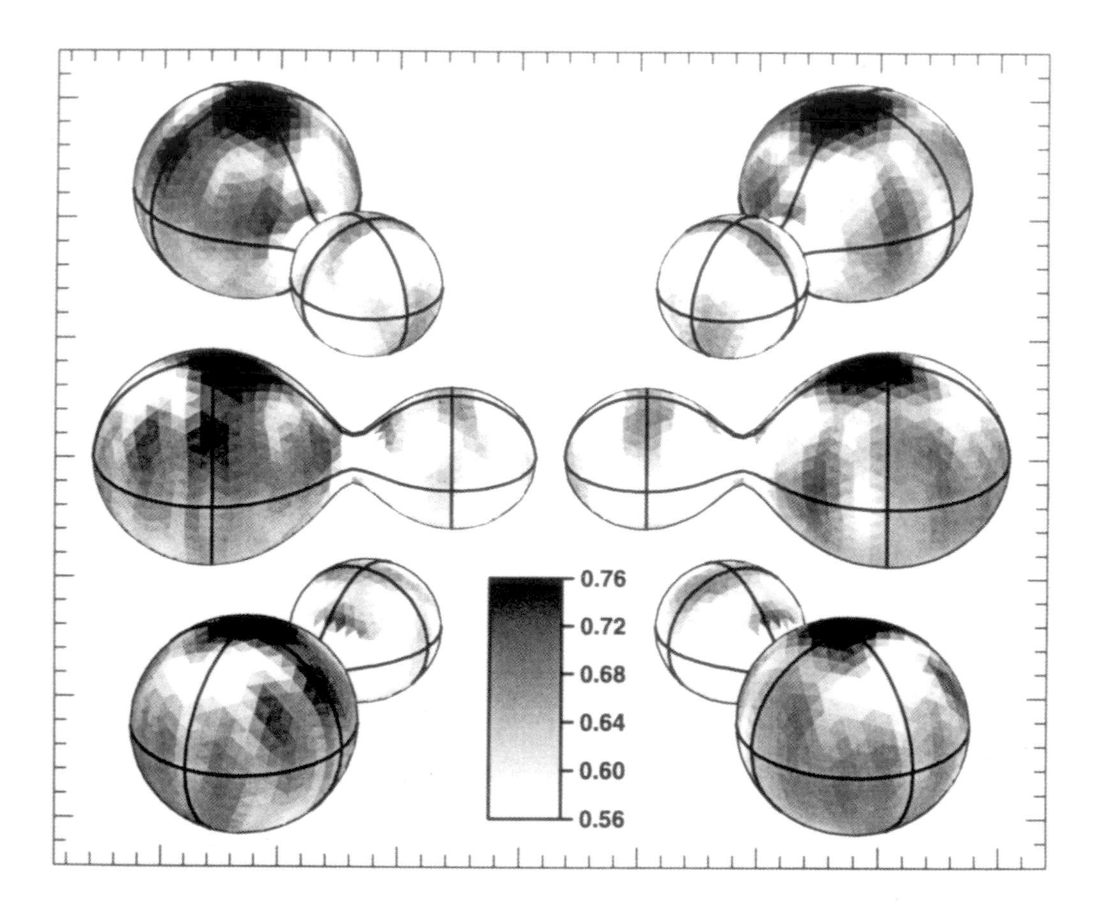

The W Ursae Majoris stars whirl around each other in actual contact, the rotation producing huge starspots. One of the breed, VW Cephei, is imaged by Doppler techniques as the stars madly orbit. *(P. D. Hendry and S. W. Mochnacki, from an article in the* Astrophysical Journal *531 [2000]: 467.)*

Residence:	URSA MAJOR
Other name:	HD 83950
Class:	F8 DWARF BINARY
Visual magnitude:	7.96
Distance:	160 LIGHT YEARS
Absolute visual magnitude:	4.48
Significance:	THE CLASSIC CONTACT BINARY.

W URSAE MAJORIS

Think "double star" and the mind jumps to classic beauties like **ALBIREO**, to **ALGOL**-type eclipsers, or perhaps to barely-resolvable stars such as **CAPELLA**. But none of these are typical; the range in doubles is too vast for any "standard" to exist. While some binaries are made of nearly identical twins, others are terribly unequal systems. Star-birth joins heavyweights with lightweights, while evolution joins white dwarfs with ordinary dwarfs, or supergiants with neutron stars and even black holes. Separations and orbital periods are just as varied. **PROXIMA CENTAURI**, connected to Alpha Centauri (itself double) by the most fragile of gravitational threads, takes perhaps 1 million years to traverse an orbit more than 10,000 AU across. Component distances range through Albireo and Capella down to systems like **RS CANUM VENATICORUM**, whose stars circle each other in less than five days. The limit is reached by W Ursae Majoris (and numerous others of its kind) where the stars are so close together that they touch! Any nearer, and they merge into one body.

Eighth magnitude W Ursae Majoris (happily shortened to "W UMa") is the visually brightest of its breed, but only because it is so nearby—just 160 light years away. Every eight hours, a pair of class F dwarfs whose centers are just 1.2 solar diameters apart, whirl around each other. The faster clips along at a speed of 225 kilometers per second (nearly ten times Earth's). Every four hours, one star moves in front of the other and produces an eclipse, cutting the visual brightness about in half. In most "normal" eclipsing binaries, one eclipse dominates the other, and outside of eclipse, the combined light is reasonably constant. Not here!

Surrounding every pair of orbiting stars is a three-dimensional figure-8 "Roche lobe." At the point between the stars where the figure-8 crosses over and the lobes meet, the effective gravity is zero, and a particle could go to either star. The components of most doubles lie well within their individual critical lobes. In many close doubles, however, one star fills its lobe and spills mass to the other. W UMa takes the concept to the limit: both stars fill these lobes and thus firmly kiss each other. From an orbiting planet, the pair would look like a giant revolving dumbbell without the connecting bar. From Earth, we continually see different aspects of the dumbbell, and as a result, the apparent brightness of W UMa smoothly changes both in and out of eclipse, the star becoming brighter, fainter, brighter, fainter, with no relaxation whatsoever. Rapid rotations, forced to be in synchrony with orbital revolution, create magnetic fields and big starspots that further complicate the light variations. Though the variations and the continually-changing spectrum are brutally difficult to analyze, they tell us that one of the F dwarfs contains one solar mass. The other only has about half, far too small for a "normal" F dwarf and surely the result of mass exchange. Both stars are about 6000 K. Strangely, the less-massive seems to be slightly warmer.

Splitting into a double is one way for a forming star to remove its spin energy. The component separations must somehow be dependent on initial conditions. By accident, or for reasons of which we are not yet aware, the W UMa pair began life very close together. Magnetic and wind interactions between the two dissipated rotational energy and drew them even closer until they touched. There seems to be only one course to follow. Becoming ever more neighborly, they will eventually merge as one rapidly spinning spotted star. **FK COMAE BERENICES**, anyone?

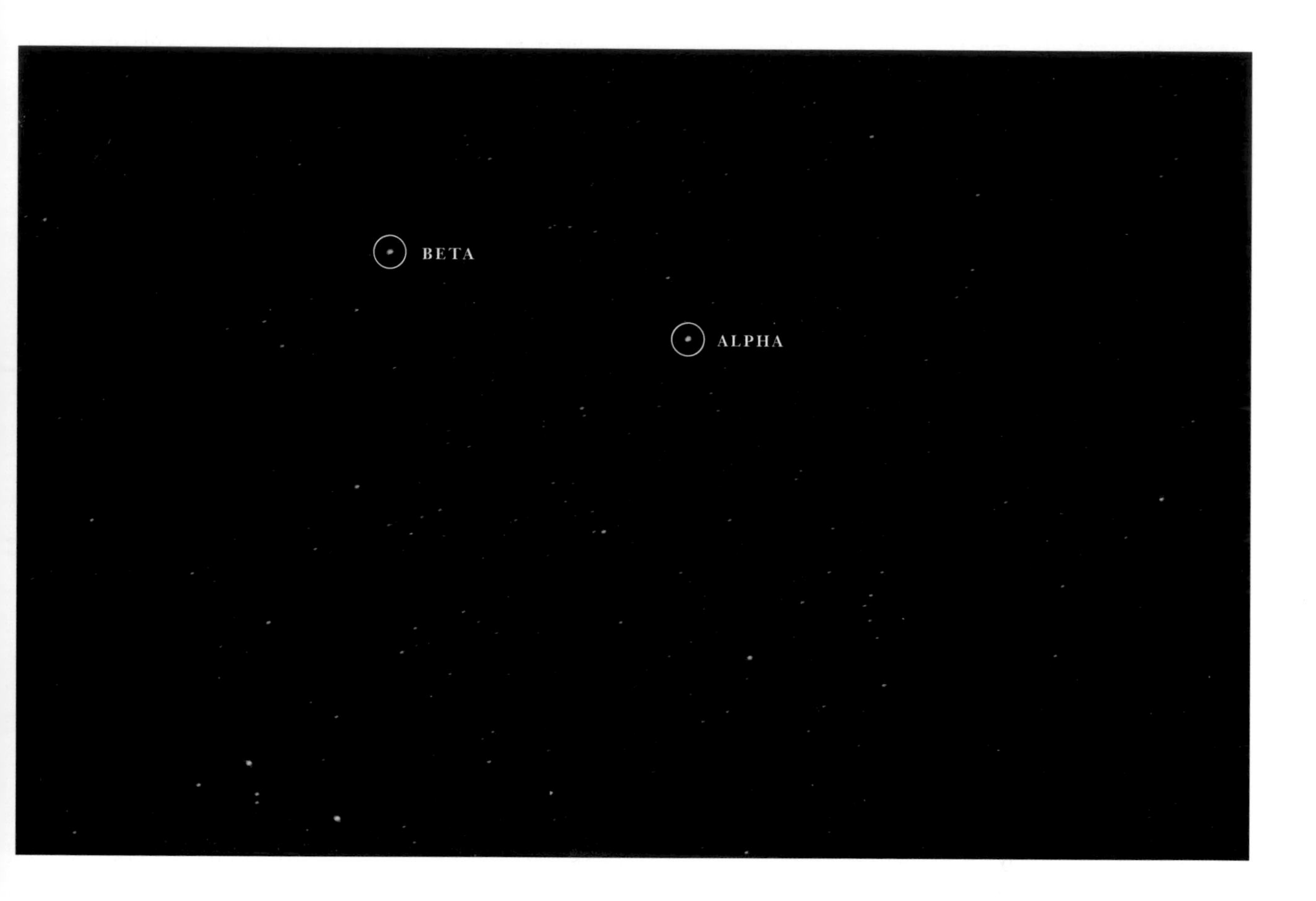

Zubenelgenubi (Alpha) and Zubeneschamali (Beta), the Scorpio's southern and northern claws, now make the constellation Libra, the Scales. *(J. B. Kaler.)*

Residence:	LIBRA
Other name:	ALPHA-2 LIBRAE
Class:	A3 SUBGIANT
Visual magnitude:	2.75
Distance:	77 LIGHT YEARS
Absolute visual magnitude:	0.88
Significance:	WITH ZUBENESCHAMALI, FORMS THE SCORPION'S CLAWS. BOTH STARS HAVE AWARD-WINNING NAMES.

ZUBENELGENUBI

To amuse a child, tell her about Zubenelgenubi, the name usually getting a giggle. Or write a planetarium show called "Zubenelgenubi and the Magic Sky," feature a talking fire hydrant and telephone pole, and send it on tour (really!). Then follow up with Zubeneschamali, and you have two of the most beloved stars in the sky. The names are among the few commonly used for stars fainter than first magnitude. Physically unrelated, the stars define the constellation Libra. Zubenelgenubi is the Alpha star and the southern of the two, while Zubeneschamali is the Beta star. The duo represent the outstretched claws of the deadly celestial scorpion, from which derive the names, Zubenelgenubi, a longish Arabic phrase meaning "the southern claw," and Zubeneschamali, referring to "the northern claw."

Zubenelgenubi is one of the heavens' few naked-eye double stars. Not quite three minutes of arc to the northwest lies a fifth magnitude class F (6000 K) dwarf called Alpha-1, rendering the brighter star Alpha-2. At the pair's distance of 77 light years, they are at least 5500 AU apart and take over 150,000 years to orbit. We know they are linked because of common movement against the distant background stars. But wait: Alpha-2 is also double. Zubenelgenubi is close enough to the ecliptic (the apparent path of the Sun) that it is regularly occulted by the Moon, which closely follows the solar path. Rather than winking out all at once, Zubenelgenubi disappears in stages, showing two stars about one hundredth of a second of arc apart. The brighter shines with around 20 solar luminosities, the fainter with around 15. Each is near two solar masses and, in spite of the spectrum, both are probably hydrogen-fusing dwarfs. The pair, at least a quarter of an AU apart, must circuit in but a few tens of days. From a hypothetical planet orbiting dimmer Alpha-1, Alpha-2 would shine about as brightly as our full Moon, the close pair 10 seconds or so seconds of arc apart and unresolvable with the Zubenelgenubian eye.

Zubeneschamali is a different beast altogether. A class B8 four-solar-mass dwarf 160 light years away (twice as far as Zubenelgenubi), it shines 300 times more brightly than the Sun (8 times more than the combined Alpha-2 pair) from a hot 11,700-K surface. Seemingly ordinary, Zubeneschamali is doubly outstanding. Stars at the cooler end of class B usually appear white to the eye. Zubeneschamali, however, has a long-standing reputation of being not just green, but deep green, nineteenth-century astronomers flowering over its hue. Heat a star and its color passes from red to orange to yellow and so on. But at a temperature where you might think green would dominate, it just comes out white. The star's color seems more to be a matter of personal perception, since only some can see it. More startling are notes from the distant past. Around 200 BC, Eratosthenes (the first to measure the size of the Earth) said the star was brighter than neighboring Antares in Scorpius. By AD 150, Ptolemy claimed that the two had equal magnitudes. Could Zubeneschamali have faded over the past couple millennia? Did Antares brighten? Stars can change. Dschubba (Delta Scorpii) recently brightened by half a magnitude and turned itself from a "normal" B star to a "Be star" like **GAMMA CASSIOPEIAE**, though the brightening hardly lasted as long as described for Zubeneschamali. The ancients clearly still have things to tell us, Zubenelgenubi's "magic sky" extending northward to encompass both of Libra's famous stars.

100

Waves on ZZ Ceti make the star vary, while waves on Earth's ocean return us home and direct us back to the SUN, where we began this journey. *(J. B. Kaler.)*

Residence:	CETUS
Other name:	ROSS 548
Class:	DA WHITE DWARF
Visual magnitude:	14.16
Distance:	150 LIGHT YEARS
Absolute visual magnitude:	10.9
Significance:	A PULSATING HYDROGEN WHITE DWARF.

ZZ CETI

A star is a gaseous ball held together by gravity, pushed outward by pressure, stirred by rotation and convection, often filled with magnetic fields. It seems amazing that most stars are relatively quiet, do not pulsate, and do not change their radii or brightnesses much. The Sun is not variable. Or is it? Are any stars perfectly stable? It all depends on how carefully one observes and on the sensitivity of the techniques. Some stars change by many magnitudes, others by one or two, some by only a tenth of or less. At what point do we stop calling them "variables"? If looked at closely enough, even our Sun really does vary, if only because sunspots move in and out of view.

Go into the middle of the range of stellar temperatures. Among the luminous F and G giants and lesser supergiants we find the pulsating Cepheids, epitomized by **DELTA CEPHEI**, which varies by nearly one magnitude over a five-day period. Drop in luminosity while climbing in temperature to the class A and F subgiants and hydrogen-fusing dwarfs, and there are the related Delta Scuti stars. Smaller and denser, these (like **BETA CASSIOPEIAE**) change by only a few hundredths of a magnitude over a period of hours in a more complex fashion, some parts of the stars moving in, others moving out.

Dimmer yet, the mid-temperature, hydrogen-rich white dwarfs (like **SIRIUS B** and **40 ERIDANI-B**, class DA) between 11,200 and 12,500 K do the same thing. These stars are so tiny and compact that if they are to vary at all, they must do so over only very short time scales. Though number two to be found, ZZ Ceti still leads the pack, its kind collectively called the "ZZ Ceti stars." From a distance of 150 or so light years, ZZ Ceti shines weakly at 14th magnitude, with a radiance only 0.7 percent that of the Sun (including an uncertain correction for ultraviolet light). Careful matching of ZZ Ceti's spectrum to mathematical models yields a temperature of 11,900 K (which places it just inside the instability limit) and a mass about 0.6 solar, typical of white dwarfs. It is the evolved carbon-oxygen core of a star that began life as a star rather like the Sun, one that has now shrunk to only twice the size of Earth.

ZZ Ceti varies with two periods at the same time, one 3 minutes 33 seconds, the other 4 minutes 34 seconds. The magnitude variations for the shorter period range from 0.002 to a bit over 0.01 (about 1 percent), and for the longer period from 0.001 to 0.008 magnitudes, about one tenth of the maximum seen for ZZ Ceti stars in general. Each period is split into two very closely-spaced intervals as a result of a 1.5-day rotation period. The pulsations are not in-and-out, but behave more like waves that run continuously around the gaseous surface of the star. Analysis suggests a lower mass around 0.54 solar.

All these pulsators have a common driving mechanism, a heat-valving layer below the surface in which atoms are becoming ionized (their electrons being stripped away). Cepheids depend on the zone in which ionized helium is becoming doubly ionized. ZZ Ceti stars are driven by a hydrogen ionization layer that is complicated by convection. Vastly smaller than Cepheids, their periods are more regular, making them among the most stable clocks known. At higher temperatures (between 21,000 and 24,000 K), the pure-helium DB white dwarfs that are epitomized by much hotter **HZ 21** similarly pulsate, driven by helium ionization instead of hydrogen. Hotter yet, we find the **PG 1159** pre-white dwarfs that are driven by carbon and oxygen ionization. In the end, no kind of star seems completely stable; it all depends on how closely we look.

Appendix A

Leading numbers are treated as if spelled out. The list includes both the principal and alternative names. ("HD" and "HR" catalogue names are not included unless they are used as a star's primary name.)

1 Acrux
2 Adhara
32 AFCRL 2688
3 AG Draconis
4 Albireo
60 Alcor
5 Algol
33 Almaaz
18 Alpha Aurigae
9 Alpha Boötis
85 Alpha Canis Majoris
17 Alpha Carinae
6 Alpha Centauri
1 Alpha Crucis
28 Alpha Cygni
94 Alpha Draconis
77 Alpha Herculis
7 Alpha Hydrae
96 Alpha Lyrae
15 Alpha Orionis
8 Alpha Scorpii
70 Alpha Ursae Minoris
88 Alpha Virginis
23 Alpha-2 Canum Venaticorum
99 Alpha-2 Librae
7 Alphard
8 Antares
9 Arcturus
10 Barnard's Star
11 Beta Canis Majoris
12 Beta Cassiopeiae
29 Beta Ceti
4 Beta Cygni
99 Beta Librae
13 Beta Lyrae
5 Beta Persei
14 Beta Pictoris
15 Betelgeuse
10 BD+04°3561
3 BD+67°122
16 The Black Widow
67 Calabash Nebula
17 Canopus
18 Capella
12 Caph
64 Cat's Eye Nebula
19 CD–38°245
53 CD–45°1841
20 CH Cygni
21 Chi Cygni
22 Chi Lupi
24 CM Tauri
23 Cor Caroli
24 Crab Nebula
52 CW Leonis
25 Cygnus X-1
26 Delta Cephei
27 Delta Orionis
28 Deneb
29 Deneb Kaitos
80 Dog Star
35 Double-Double in Lyra
30 DQ Herculis
34 18 Eridani
60 80 Ursae Majoris
31 EG 129
32 Egg Nebula
44 Eltanin
33 Epsilon Aurigae
2 Epsilon Canis Majoris
34 Epsilon Eridani
35 Epsilon Lyrae
36 ESO 439–26
37 Eta Carinae
38 FG Sagittae
39 51 Pegasi
22 5 Lupi
79 5 Ophiuchi
40 FK Comae Berenices
41 40 Eridani
42 FU Orionis
47 G″
43 Gamma Cassiopeiae
44 Gamma Draconis
45 Gamma-2 Velorum
83 GB 790324
46 GD 165B
47 Geminga
48 Gliese 229B
49 GRS 1915+105
69 GW Virginis
50 HD 93129 A
38 Henize 1-5
61 Herschel's Garnet Star
75 Hind's Crimson Star

FULL NAME LIST

76 Hubble's Variable Nebula
71 Hulse-Taylor binary pulsar
51 HZ 21
55 IRAS 04 287+1801
52 IRC +10 216
53 Kapteyn's Star
41 Keid
54 Kepler's Star
55 L 1551 IRS 5
56 Lambda Boötis
57 LB 11146
31 LHS 3424
58 Merope
27 Mintaka
59 Mira
11 Mirzam
60 Mizar
61 Mu Cephei
62 Mu Columbae
63 MXB 1730-335
64 NGC 6543
65 NGC 7027
37 Nova Carinae 1843
68 Nova Cygni 1600
66 Nova Cygni 1975
30 Nova Herculis 1934
82 Nova Ophiuchi
42 Nova Orionis 1939
24 NP 0532
67 OH 231.8 +4.2
59 Omicron Ceti
41 Omicron-2 Eridani
68 P Cygni
69 PG 1159-035
65 PK 084–03 1
70 Polaris
84 Polaris Australis
6 Proxima Centauri
24 PSR B0531+21.9
71 PSR B1257+12
72 PSR B1913+16
16 PSR B1957+20
67 QX Puppis
16 QX Sagittae
73 R Aquarii
74 R Coronae Borealis
75 R Leporis
76 R Monocerotis
63 Rapid Burster
77 Rasalgethi
45 Regor
78 Rho Cassiopeiae
79 Rho Ophiuchi
6 Rigil Kentaurus
100 Ross 548
80 RR Lyrae
81 RS Canum Venaticorum
82 RS Ophiuchi
91 Sanduleak –69°202
78 7 Cassiopeiae
83 SGR 1900+14
13 Sheliak
84 Sigma Octantis
85 Sirius
86 16 Cygni
87 61 Cygni
88 Spica
89 SS Cygni
90 SS 433
0 The Sun
91 Supernova 1987A
24 Supernova of 1054
95 Supernova of 1572
54 Supernova of 1604
92 T Tauri
93 Theta-1 Orionis
94 Thuban
93 Trapezium
34 18 Eridani
43 27 Cassiopeiae
26 27 Cephei
58 23 Tauri
95 Tycho's Star
6 V 645 Centauri
54 V 843 Ophiuchi
58 V 971 Tauri
90 V 1343 Aquilae
25 V 1357 Cygni
49 V 1487 Aquilae
66 V 1500 Cygni
96 Vega
97 VV Cephei
98 W Ursae Majoris
57 WD 0945+245
51 WD 1211+332
60 Zeta Ursae Majoris
99 Zubenelgenubi
99 Zubeneschamali
100 ZZ Ceti

Appendix B

Use this guide to follow the course of stellar evolution from birth through stable main-sequence life to the various forms of stellar death. Dwarfs are aranged from cool to hot; within other categories, the stars are in no particular order. In the case of doubles, they have been placed in the category of the star that is of greater significance to the evolution of the pair.

PRE-MAIN SEQUENCE

55 L 1551 IRS 5
92 T Tauri
42 FU Orionis
76 R Monocerotis

MAIN SEQUENCE (DWARFS)

48 Gliese 229 B
46 GD 165B
6 Proxima Centauri
10 Barnard's Star
53 Kapteyn's Star
87 61 Cygni
34 Epsilon Eridani
98 W Ursae Majoris
6 Alpha Centauri
0 Sun
39 51 Pegasi
86 16 Cygni
81 RS Canum Venaticorum
14 Beta Pictoris
35 Epsilon Lyrae
99 Zubenelgenubi
85 Sirius
60 Mizar (and Alcor)
23 Cor Caroli
56 Lambda Boötis
96 Vega
99 Zubeneschamali
05 Algol
88 Spica
2 Adhara
62 Mu Columbae
93 Theta-1 Orionis

SUBGIANTS

5 Algol
12 Beta Cassiopeiae
79 Rho Ophiuchi
43 Gamma Cassiopeiae
22 Chi Lupi
58 Merope
1 Acrux

FIRST-ASCENT AND HELIUM-BURNING GIANTS

84 Sigma Octantis
19 CD–38°245
9 Arcturus
4 Albireo
18 Capella
17 Canopus
7 Alphard
26 Delta Cephei
80 RR Lyrae
94 Thuban
70 Polaris
27 Delta Orionis
11 Beta Canis Majoris
13 Beta Lyrae
29 Deneb Kaitos
40 FK Comae Berenices
3 AG Draconis
82 RS Ophiuchi

SECOND-ASCENT GIANTS

44 Gamma Draconis
59 Mira
21 Chi Cygni
20 CH Cygni
73 R Aquarii
75 R Leporis
52 IRC +10 216
67 OH 231.8 +4.2

PLANETARY NEBULAE AND RELATED OBJECTS

32 Egg Nebula
64 NGC 6543
65 NGC 7027
69 PG 1159–035
38 FG Sagittae
74 R Coronae Borealis

WHITE DWARFS

41 40 Eridani
85 Sirius B
31 EG 129
36 ESO 439-26
51 HZ 21

STARS LISTED BY EVOLUTION

100 ZZ Ceti
57 LB 11146
30 DQ Herculis
66 Nova Cygni 1975
89 SS Cygni

SUPERGIANTS

77 Rasalgethi
8 Antares
15 Betelgeuse
28 Deneb
33 Epsilon Aurigae
61 Mu Cephei
97 VV Cephei
50 HD 93129A
78 Rho Cassiopeiae
68 P Cygni
37 Eta Carinae
45 Gamma-2 Velorum

SUPERNOVAE

95 Tycho's Star
54 Kepler's Star
91 Supernova 1987A

NEUTRON STARS

24 Crab Nebula
47 Geminga
63 MXB 1730–335
16 Black Widow
71 PSR B1257+12
83 SGR 1900+14
90 SS 433
72 PSR B1913+16

BLACK HOLES

25 Cygnus X-1
49 GRS 1915+105

Appendix C

Declination and Right Ascension are the celestial analogues to terrestrial latitude and longitude. Declination, measured in standard degrees, °, minutes (′, each a 60th of a degree), and seconds (″, each a 60th of a minute) is the angular distance of the star north (+) or south (–) of the celestial meridian. Right ascension, measured in time units of hours (h), minutes (m), and seconds (s), is the angular distance of the star eastward of the Vernal Equinox in Pisces. The right ascension is the time it takes for the sky to rotate from the crossing of the Vernal Equinox over the celestial meridian (the north–south line) to the crossing of the star (in sidereal time units).

	STAR	RIGHT ASC.	DEC.
		h m s	° ′ ″
0	The Sun	– – –	– – –
12	Beta Cassiopeiae	00 09 10.7	+59 08 59
95	Tycho's Star	00 25 20.0	+64 08 19
29	Deneb Kaitos	00 43 35.4	–17 59 12
19	CD–38 245	00 46 36.2	–37 39 34
43	Gamma Cassiopeiae	00 56 42.5	+60 43 00
100	ZZ Ceti	01 36 11	–11 20 48
59	Mira	02 19 20.8	–02 58 40
70	Polaris	02 31 49.1	+89 15 51
5	Algol	03 08 10.1	+40 57 20
34	Epsilon Eridani	03 32 55.8	–09 27 30
58	Merope	03 46 19.6	+23 56 54
41	40 Eridani	04 15 16.3	–07 39 10
92	T Tauri	04 21 59.4	+19 32 06
55	L 1551 IRS 5	04 31 33.6	+18 08 15
75	R Leporis	04 59 36.3	–14 48 23
33	Epsilon Aurigae	05 01 58.1	+43 49 24
53	Kapteyn's Star	05 11 40.6	–45 01 06
18	Capella	05 16 41.4	+45 59 53
27	Delta Orionis	05 32 00.4	–00 17 57
24	Crab Nebula	05 34 32.0	+22 00 52
93	Theta-1 Orionis	05 35 15.1	–05 23 29
91	Supernova 1987A	05 35 28.3	–69 16 13
42	FU Orionis	05 45 22.6	+09 14 12
62	Mu Columbae	05 45 59.9	–32 18 23
14	Beta Pictoris	05 47 17.1	–51 04 00
15	Betelgeuse	05 55 10.3	+07 24 25
48	Gliese 229B	06 10 35.1	–21 51 18
11	Beta Canis Majoris	06 22 42.0	–17 57 21
17	Canopus	06 23 57.1	–52 41 44
47	Geminga	06 33 54.2	+17 46 12
76	R Monocerotis	06 39 09.8	+08 44 12
85	Sirius	06 45 08.9	–16 42 58
2	Adhara	06 58 37.5	–28 58 20
67	OH 231.8 +4.2	07 42 16.8	–14 42 52
45	Gamma-2 Velorum	08 09 32.0	–47 20 12
7	Alphard	09 27 35.2	–08 39 31
98	W Ursae Majoris	09 43 45.5	+55 57 09
52	IRC +10 216	09 47 57.3	+13 16 44
57	LB 11146	09 48 46.7	+21 21 30
50	HD 93129 A	10 43 57.5	–59 32 51
37	Eta Carinae	10 45 03.6	–59 41 04
36	ESO 439–26	11 39 03.6	–28 52 18
69	PG 1159–035	12 01 46.0	–03 45 39
51	HZ 21	12 13 56.3	+32 56 31
1	Acrux	12 26 36.2	–63 05 57
23	Cor Caroli	12 56 01.7	+38 19 06
71	PSR B1257+12	13 00 01	+12 40 00
81	RS Canum Venaticorum	13 10 36.9	+35 56 06
60	Mizar	13 23 55.2	+54 55 25
88	Spica	13 25 11.6	–11 09 41
60	Alcor	13 25 13.5	+54 59 17

STARS BY POSITION

40	FK Comae Berenices	13 30 46.8	+24 13 58
94	Thuban	14 04 23.3	+64 22 33
9	Arcturus	14 15 39.7	+19 10 57
56	Lambda Boötis	14 16 23.0	+46 05 18
46	GD 165B	14 24 40	+09 17 24
6	Proxima Centauri	14 29 42.9	–62 40 46
6	Alpha Centauri	14 39 36.2	–60 50 08
99	Zubenelgenubi	14 50 52.7	–16 02 30
99	Zubeneschamali	15 17 00.4	–09 22 59
74	R Coronae Borealis	15 48 34.4	+28 09 24
22	Chi Lupi	15 50 57.5	–33 37 38
3	AG Draconis	16 01 41.0	+66 48 10
79	Rho Ophiuchi	16 25 35.1	–23 26 50
8	Antares	16 29 24.5	–26 25 55
77	Rasalgethi	17 14 38.9	+14 23 25
54	Kepler's Star	17 30 37.2	–21 28 51
63	MXB 1730–335	17 33 24.1	–33 23 16
82	RS Ophiuchi	17 50 13.2	–06 42 28
44	Gamma Draconis	17 56 36.4	+51 29 20
10	Barnard's Star	17 57 56.9	+04 14 52
64	NGC 6543	17 58 33.4	+66 38 00
30	DQ Herculis	18 07 30.2	+45 51 32
96	Vega	18 36 56.3	+38 47 01
35	Epsilon Lyrae	18 44 21.5	+39 36 39
13	Beta Lyrae	18 50 04.8	+33 21 46
31	EG 129	19 00 10	+70 39 36
83	SGR 1900+14	19 07 14.2	+09 19 19
49	GRS 1915+105	19 15.11.5	+10 56 44
72	PSR B1913+16	19 15 28.1	+16 06 28
20	CH Cygni	19 24 33.1	+50 14 29
80	RR Lyrae	19 25 27.9	+42 47 04
4	Albireo	19 30 45.4	+27 57 55
86	16 Cygni	19 41 49.0	+50 31 30
21	Chi Cygni	19 50 33.9	+32 54 51
90	SS 433	19 51 48.6	–05 48 47
25	Cygnus X-1	19 58 21.7	+35 12 06
16	Black Widow	19 59 36.8	+20 48 15
38	FG Sagittae	20 11 56.1	+20 20 04
68	P Cygni	20 17 47.2	+38 01 59
28	Deneb	20 41 25.9	+45 16 49
32	Egg Nebula	21 02 18.8	+36 41 38
87	61 Cygni	21 06 54.6	+38 44 45
65	NGC 7027	21 07 01.7	+42 14 10
84	Sigma Octantis	21 08 46.8	–88 57 23
66	Nova Cygni 1975	21 11 36.6	+48 09 02
89	SS Cygni	21 42 42.7	+43 35 10
61	Mu Cephei	21 43 30.5	+58 46 48
97	VV Cephei	21 56 39.1	+63 37 32
26	Delta Cephei	22 29 10.3	+58 24 55
39	51 Pegasi	22 57 28.0	+20 46 08
73	R Aquarii	23 43 49.5	–15 17 04
78	Rho Cassiopeiae	23 54 23.0	+57 29 58

Glossary

Cross references to other entries in the Glossary are in *italics.*

A STARS White *stars* with the strongest hydrogen absorptions; the *dwarfs* fall between 7200 and 9500 K.
ABSOLUTE VISUAL MAGNITUDE The *apparent magnitude* a *star* would have at a distance of 10 *parsecs*, or 32.6 *light years*.
ABSORPTION LINES Narrow absorptions in a stellar *spectrum*; produced by the absorption of energy at specific wavelengths by the *electrons* bound to all the various kinds of atoms, *ions*, and molecules.
APPARENT MAGNITUDE The *magnitude* of a *star* as seen from Earth.
APPARENT VISUAL MAGNITUDE The *apparent magnitude* of a *star* as seen by the human eye (or by a similar yellow-green detector).
ASTRONOMICAL UNIT The average distance between Earth and Sun, 150 million kilometers, and the fundamental distance unit in astronomy. The "*AU.*"
AU The *Astronomical Unit.*
B STARS High temperature blue-white *stars* defined by hydrogen and neutral helium absorptions; the *dwarfs* fall between 10,000 and 30,000 K.
BINARY STARS *Double stars.*
BLACK HOLES The by-products of *supernovae* that have collapsed past the spherical radius (the *event horizon*) at which the escape velocity equals the speed of light, rendering black holes intrinsically dark.
BROWN DWARFS Substars fainter and cooler than ordinary main sequence *dwarfs;* brown dwarfs are too cool inside to fuse hydrogen (as hydrogen-1), but they can fuse their natural *deuterium* (hydrogen-2). They are all *L stars* or *T stars.*
C STARS *Carbon stars.*
CARBON CYCLE The process by which hydrogen is fused to helium using carbon as a nuclear catalyst. Operates in higher-mass *stars.*
CARBON STARS *Second-ascent giant stars* in which carbon dominates over oxygen.
DEUTERIUM Heavy hydrogen; hydrogen whose nucleus is made of a *proton* attached to a *neutron*; hydrogen-2.
DOPPLER EFFECT The apparent shift in the wavelength of a wave as a result of motion along the line of sight. *Stars* exhibit the effect through wavelength shifts in their *absorption lines.*
DOUBLE STARS *Stars* that have two components, two separate stars, in orbit about each other; *binaries*; used to determine stellar masses.
DWARF STARS Ordinary core hydrogen-burning *stars* of the *main sequence*, OBAFGKML, which define a mass sequence from 100 solar masses through 0.08 solar masses, below which we find the *brown dwarfs.*
ECLIPSING BINARY A *double star* in which the orbital plane is nearly in the line of sight, so that the two stars alternately eclipse each other.
ELECTROMAGNETIC RADIATION Radiation that includes light and all other parts of the *electromagnetic spectrum*, consisting of waves of alternating magnetic and electric fields and that travel at the speed of light, 300,000 kilometers per second.
ELECTROMAGNETIC SPECTRUM The full array of *electromagnetic radiation*, which extends from *gamma rays* to *radio.*
ELECTRONS Negatively-charged particles that surround atomic nuclei.

EMISSION LINES Narrow bright emissions in the spectra of interstellar clouds or in *stars*; produced by the emission of energy at specific wavelengths by the electrons bound to all the various kinds of atoms and *ions*; the reversal of absorption lines.

EVENT HORIZON The spherical "surface" of a *black hole* at which the escape velocity equals the speed of light.

EVOLUTION See *stellar evolution*.

F STARS Intermediate-temperature yellow-white *stars* (6000 - 7000 K for *dwarfs*) defined by strong hydrogen and by ionized and neutral metals.

FIRST-ASCENT GIANT STARS *Stars* brightening as *giants* for the first time, with dead helium cores.

G STARS Solar-temperature yellowish *stars* (the *dwarfs* falling between 5300 and 5900 K) characterized by weakening ionic spectra and strengthening neutral spectra.

GALAXY Our "local" collection of stars, 80,000 *light years* across. Our Galaxy is comprised of a populous disk of 200 billion *stars*, the whole affair surrounded by a sparse, extensive, ancient, metal-deficient halo some 13 billion years old.

GAMMA RAYS The shortest-wavelength and highest-energy *electromagnetic radiation*.

GIANT STARS Luminous evolved *stars* whose cores have gone past hydrogen burning; most prominent as class K and M red giants. *First-ascent giants* have dead helium cores, while *second-ascent giants* have dead carbon cores.

HALO The sparsely-populated spherical realm of *stars* that surrounds the *Galaxy*'s disk.

HYPERGIANTS The most luminous of all evolved *stars*.

INFRARED Lower-energy radiation with wavelengths longer than the eye can see.

INTERSTELLAR MEDIUM The dusty gases of interstellar space.

IONS Atoms with one or more *electrons* missing.

K DEGREES *Kelvin degrees*.

KELVIN DEGREES A measure of temperature in Celsius (centigrade) degrees above absolute zero (which is –273°C).

K STARS Lower-temperature orange *stars* (*dwarfs* falling between 4000 and 5200 K) with strong *absorption lines* of neutral metals.

L STARS The lowest temperature hydrogen-fusing *dwarfs*, red to infrared *stars* with temperatures between 1500 and 2000 K, defined by neutral metals and metallic hydrides; about half are *brown dwarfs*.

LIGHT YEAR The distance light will travel (at 300,000 kilometers per second) in 1 year (of 31 million seconds). A light year = 0.307 *parsecs* = 63,240 *AU*.

LUMINOSITY The total power output of a *star* in watts or in units of the solar luminosity.

M STARS Low-temperature orange-red *stars* (2000 to 3900 K for *dwarfs*) recognized by absorption bands of molecules (mostly titanium oxide); the coolest class for *giants* and *supergiants*.

MAGNITUDE A measure of the apparent or absolute brightness of a star, wherein 5 magnitudes corresponds to a factor of 100 in brightness, 1 magnitude to a factor of 2.512...

MAIN SEQUENCE STARS Ordinary core hydrogen (hydrogen-1) burning *dwarfs*, OBAFGKML, defining a mass sequence from 100 solar masses through 0.08 solar masses, below which we find the *brown dwarfs*.

MASS-LUMINOSITY RELATION The positive relation between the masses of *dwarf stars* and their *luminosities*.

MILKY WAY The band of light around the sky caused by the combined light of the disk of our *Galaxy*.

N STARS Cool *carbon stars*, *second-ascent* deep red *giants* whose temperatures parallel those of *M stars*.
NEUTRINOS Nearly-massless particles that are produced by atomic reactions and that carry away energy.
NEUTRON-STAR LIMIT The maximum mass, about 2 or 3 solar masses, beyond which a *neutron star* will collapse.
NEUTRON STARS The dense remains of iron-core-collapse *supernovae*, made (obviously) of *neutrons*.
NEUTRONS Neutral particles with about the mass of a *proton* that help compose atomic nuclei.
O STARS High-temperature (32,000 to 50,000 K for *dwarfs*), high-mass (greater than 12 solar) blue *stars* characterized by hydrogen and by neutral and ionized helium absorptions. O stars explode as *supernovae*.
PARALLAX Half the angle through which a *star* appears to shift as the Earth goes about the Sun, from which distance is derived. The distance in *parsecs* is the inverse of the parallax.
PARSEC A professionally-used distance unit. At one parsec, the Earth's orbital radius appears 1 second of arc across. Distance in parsecs is the inverse of the *parallax* in seconds of arc. One parsec = 206,265 *AU* = 3.26 *light years*.
PLANETARY NEBULAE The (mis-named) ejected outer envelopes of *second-ascent giant stars*, ionized and illuminated by the hot, nearly exposed old nuclear-burning cores.
PROTON-PROTON CHAIN The process by which four hydrogen atoms (*protons*) accumulate by successive collisions to form helium.
PROTONS The positively-charged particles that constitute the atomic nucleus; a hydrogen nucleus. The number of protons in the nucleus determines the kind of chemical element.
PULSARS Rapidly spinning *neutron stars* that beam energy out along a tilted magnetic field.
R STARS Warmer *carbon stars*, *second-ascent* yellow-orange *giants* whose temperatures parallel those of the of *K* and cooler *G stars*.
RADIO *Electromagnetic radiation* of the longest wavelengths, extending longward from about 1 millimeter.
S STARS Transition *carbon stars*, *second-ascent* red *giants* whose temperatures parallel those of *M* and *N stars*, but for which the abundance of carbon equals that of oxygen; characterized by absorption bands of zirconium oxide.
SECOND-ASCENT GIANT STARS *Stars* brightening as *giants* for the second time, with dead carbon-oxygen cores.
SPECTRAL CLASS The class of a *star* within the *spectral sequence*, OBAFGKMLT (or, as combined with chemical classes, R, N, and S).
SPECTRAL SEQUENCE: *Stars* arranged by temperature according to the appearance of their spectra: OBAFGKMLT. See the spectral sequence on page XVIII.
SPECTROSCOPIC BINARIES *Double stars* whose components are so close together that they can generally be separated only by the *Doppler effect* on their *spectra*.
SPECTRUM Visually, the array of colors of light from red through violet, or the array of the lengths of the waves that make them. The spectrum extends to longer wavelengths through the *infrared* to *radio* and to shorter wavelengths through the *ultraviolet* to the *X-* and *gamma ray* spectral domains.
STAR A body that is undergoing thermonuclear fusion, or that has spent or will spend time in such a state.

STELLAR EVOLUTION The processes that involve the aging of a *star* from birth through death.

SUBDWARFS Ancient *dwarfs* of lower *spectral classes* that have lower metal contents and more properly belong to the *Galaxy*'s halo.

SUBGIANTS *Stars* that have given up (or are about to give up) core-hydrogen fusion and are in the process of becoming *giants*.

SUPERGIANTS High-mass luminous *stars* that have evolved from *O stars* and hot *B stars*; most prominent as *class M* red supergiants; the largest of all stars.

SUPERNOVA REMNANT The outwardly exploded remains of a *supernova*, which may be the stellar matter itself or the outbound shock wave acting on the gases of the *interstellar medium*.

SUPERNOVAE Exploding *stars* with two origins: high-mass evolved *stars* of (mostly) *class O* whose iron cores collapse, and *white dwarfs* that accrete matter from companions and that overflow the *white-dwarf limit* of 1.4 solar masses.

T STARS The coolest *stars*, defined by methane absorption bands, with temperatures around 1000 K; these infrared stars are all *brown dwarfs*.

ULTRAVIOLET High-energy radiation with wavelengths just shorter than the eye can see.

VISUAL BINARIES *Double stars* in which the components can be separated at the telescope by eye.

WHITE-DWARF LIMIT The maximum mass, 1.4 solar masses, allowed for a *white dwarf*, beyond which it will collapse.

WHITE-DWARF SUPERNOVAE *Supernovae* caused by *white dwarfs* that accrete matter from companions and overflow the *white-dwarf limit*.

WHITE DWARFS The end-products (the dead carbon-oxygen cores) of the middle part of the main sequence that lie between 0.8 and 10 solar masses (*spectral class* G8 to B2).

X-RAY Short, high-energy *electromagnetic radiation* that falls between the *ultraviolet* and the *gamma ray* spectrum.